"GRASS ROOTS":

Perspectives of Senior Non-Commissioned Officers on Operations

IN HARM'S WAY

"GRASS ROOTS": Perspectives of Senior Non-Commissioned Officers on Operations

Edited by
Dr. Emily Spencer

CANADIAN DEFENCE ACADEMY PRESS

Canadian Defence Academy Press
PO Box 17000 Stn Forces
Kingston, Ontario K7K 7B4

Produced for the Canadian Defence Academy Press
by 17 Wing Winnipeg Publishing Office.
WPO30333

Cover Photo:
Warrant Officer Ray MacFarlane discusses tactical response with Afghan National Army forces during a troops in contact (TIC) in Panjwayi. Photo by B. Horn.

Library and Archives Canada Cataloguing in Publication

Grassroots : perspectives of senior non-commissioned officers on operations / edited by Emily Spencer.

(In harm's way)
Includes bibliographical references and index.
Issued by Canadian Defence Academy.
ISBN 978-0-662-47813-3 (bound).--ISBN 978-0-662-47814-0 (pbk)
Cat. no.: D4-4/4-1-2008E (bound) -- Cat. no.: D4-4/4-2-2008E (pbk)

1. Canada--Armed Forces--History--20th century. 2. Canada--Armed Forces--Foreign countries. 3. Military art and science--Canada. 4. Canada--Armed Forces-Non-commissioned officers. 5. Operational art (Military science). 6. Command of troops. I. Spencer, Emily II. Canadian Defence Academy III. Title: Perspectives of senior non-commissioned officers on operations. IV. Series.

U163.G72 2008 355.4'771 C2008-980051-6

Printed in Canada.

3 5 7 9 10 8 6 4 2

ACKNOWLEDGEMENTS

I would be remiss if I did not acknowledge the hard work and efforts of all those who made this volume possible. First, I would like to thank the contributors for their dedication and sense of duty. This volume was possible only because of the invaluable work that they do on operations and the time and effort that they put in afterwards in re-telling their experiences. Although all leaders are responsible for preparing their successors for the challenges that they may face in the future, many fail to do so. These contributors, however, in their off-duty hours, dedicated themselves to passing down their leadership experiences. Their efforts are greatly appreciated.

I would also like to thank the Acting Director of the Canadian Forces Leadership Institute, Commander Bob Edwards, for bringing me on board such an interesting and valuable project. Equally, I would like to thank the Chair of the Canadian Defence Academy Press, Colonel Bernd Horn, for his constant tutelage and his dedication to excellence.

Last, but not least, I must express my sincere gratitude to the men and women who serve in the Canadian Forces. It is their sacrifices and achievements that have made the *In Harm's Way* series a success.

TABLE OF CONTENTS

Foreword iii

Preface v

Introduction Crucible of Success: The Senior Non-Commissioned Officer Corps
Bernd Horn ix

Chapter 1 Building for the Future: the Construction of Camp Mirage, Operation Apollo, Rotation Zero
Charles Malchow 1

Chapter 2 Walking The Line and Keeping Them Happy: The Leadership Experiences of CPO1 Eric Carberry on Operation Apollo, Rotation Zero
Craig Leslie Mantle 13

Chapter 3 Operation Apollo through the Eyes of the Battle Group RSM, February-July 2002
J.A. Comeau 21

Chapter 4 Leadership Lessons Learned
Alex Bouzane 39

Chapter 5 "Loose Lips Sink Ships": The Leadership Experiences of CPO1 Ken Mullins on Operation Apollo, Rotation 2
Craig Leslie Mantle 55

Chapter 6 The Candyman Cometh: The Leadership Experiences of CPO1 (Ret'd) Pat Evans on Operation Apollo, Rotation 3
Craig Leslie Mantle 65

Chapter 7 Lessons Learned from a Camp Sergeant-Major Operation Athena, Rotation Zero, Kabul, Afghanistan
T.J.P. Mugford 79

Chapter 8 Leadership Challenges during Formed and Composite Unit Deployments
Thomas J. Secretan 83

Chapter 9 Setting Standards for the Future: Advice from an RSM
André Jutras 107

Chapter 10 Establishing a Canadian Footprint in Afghanistan: Insight from an RSM during Operation Athena
Michael Poirier 115

Chapter 11 Combat Tour in Afghanistan: Lessons Learned from an RSM
Randy Northrup 121

Chapter 12 Words of Wisdom from a Task Force RSM: Operation Athena, Rotation 3 – February-August 2005
Lee Topp 131

Chapter 13 Task Force Afghanistan 2006
Kit Charlebois 165

Chapter 14 Animosity in the Ranks: the Difficulty in Integrating Reserve Augmentees into Regular Force Battalions
Kurt Grant 175

Contributors 189

Glossary 193

Index 199

FOREWORD

I am delighted to introduce *"Grass Roots": Perspectives Of Senior Non-Commissioned Officers On Operations*. This book represents the fourth volume of the seminal *In Harm's Way* series created by the Canadian Forces Leadership Institute (CFLI) and published by the CDA Press which has over 30 titles now in print. As such, the Canadian Defence Academy is well on its way to capturing operational experience that can be used in our professional development institutions. After all, this book, like those before it and those still in press, are an integral component of our Strategic Leadership Writing Project, which is designed to (a) create a distinct and unique body of Canadian leadership literature and knowledge that will assist leaders at all levels of the Canadian Forces in preparing themselves for operations in a complex security environment, and (b) inform the public with respect to the contribution of Canadian Forces service personnel to Canadian society and international affairs.

This volume, as well as those that preceded it and those that will follow it in the *In Harm's Way* series, represents the experiences and insights of an array of individuals who have taken the time and effort to capture their thoughts so that others may profit from them. Invariably, this comes at the sacrifice of their well deserved personal and family time to complete their stories. Their commitment and dedication is greatly appreciated. I wish to thank all those who have risen to the challenge of preparing our successors for the complexities that lie ahead.

"Grass Roots" is an important addition to the series. Similar to the other volumes, it contains a collection of "war stories" from all three environments and from a myriad of different operations. They are based on personal experiences and the respective interpretations, reflections and lessons learned. Importantly, they are the experiences of our Senior NCO Corps, who have always represented the core of any military institution. As the first layers in the chain of command they conduct the daily face-to-face "business" of our profession whether in garrison, on exercise or on operations domestically or around the globe. They are responsible for training, disciplining and leading the men and women of the Canadian Forces (CF) and deserve the credit and recognition for allowing the CF to remain an effective and credible institution at home and abroad. Although the experiences described in this book do not represent the doctrine and

policy of the Department of National Defence or of the Canadian Forces, they are no less valid. They express a richness of information that can assist others in preparing for operations and leadership in general. In essence, they can act as virtual experience for those who have not had the opportunity to deploy. However, even those with operational experience can profit from a wider, broader and greater repertoire of knowledge.

In closing, I wish to re-iterate the importance of this book. I hope it will provide valuable insight for those who serve in, and for those who interact with, the profession of arms in Canada.

Chief-Warrant Officer Greg Lacroix
Canadian Forces Chief Warrant Officer

PREFACE

Perspectives of Senior Non-Commissioned Officers on Operations is the fourth volume in the Canadian Forces Leadership Institute's seminal *In Harm's Way* series. This book continues the examination of the experiences and lessons learned by Canadian service personnel while serving on operations at home and abroad. As with the first three volumes, its strength lies in the experiences of the contributors. Specifically, this volume provides valuable insight from senior non-commissioned officers from all three services spanning the last two decades.

Continuing with a common theme in the series, it must be re-iterated that the dedication, initiative and professionalism of the men and women in the Canadian Forces is second to none. They have consistently done their country proud, persevering under harsh conditions generally in remote locations and often in dangerous and volatile conditions. They have faced ambiguity, change and uncertainty with courage and conviction – often while enduring shortages in manpower and equipment. Nonetheless, they have remained humble about their achievements.

While exceptionally commendable, this undercurrent of humility is not without vulnerability. First, this silence fails to educate Canadian society in regard to the contributions and sacrifices of its military personnel. Consequently, Canadians, including political leaders, are not always fully cognisant of the difficulties achieving, or the significance of, CF personnel's achievements.

More important, however, the reluctance to share operational experiences deprives the institution of invaluable learning opportunities. Although the security environment is dynamic and each deployment and theatre of operation is unique, there are still many lessons that can be learned through the experiences of others. Leadership challenges, in particular, provide a plethora of professional development opportunities. There are many situations, problems and dilemmas that are timeless and not particular to mission or geographic area. The sharing of these challenges and/or solutions by individuals provides guidance to others, regardless of rank or seniority. Additionally, lessons from veterans can provide insight for members serving on future missions, wherever they may be, that will allow them to be better prepared to meet the challenges and to lead their personnel more effectively.

The chapters in *Perspectives of Senior Non-Commissioned Officers on Operations* also provide a glimpse into the different cultures of the various services (Air, Army and Navy) in addition to a view of the multitude of problems that confront individuals in the various classifications and trades of the military occupations. This insight will assist all military personnel, as well as the Canadian public, in developing a clear understanding of the peculiar and distinct challenges that individuals in the military face. This understanding is critical as we advance further into the new and dynamic defence environment of the 21st Century.

Overall, this sharing of knowledge and experience throughout the institution is key to enabling mission success. It is also the keystone of a learning organization. For this reason, the Canadian Forces Leadership Institute established the Strategic Leadership Writing Project, a seminal programme intended to create a distinct and unique body of Canadian leadership literature and knowledge that will assist leaders at all CF rank levels in preparing for operations in a complex security environment. The project will also help to educate the public regarding the vast contributions and sacrifices by CF personnel to Canadian society and international affairs.

The *In Harm's Way* series, in essence, is a series of first hand accounts by CF members on operations that capture the challenges faced by leaders at all rank levels and from all three services on operations dating from the early post-Cold War period of the 1990s to the present. These "war stories" are intended to act as a professional development tool for military members as well as educate the Canadian public.

The chapters contained in *Perspectives of Senior Non-Commissioned Officers on Operations* are the experiences and viewpoints of the authors and should be taken as such. Personal attitudes, biases, beliefs and interpretations are clearly at play. Readers may not agree with all the views, opinions or statements that appear on the following pages. Indeed, what is contained in this book is not meant to represent CF doctrine or official policy. Rather, it is meant to showcase the personal experiences and reflections of individuals who have been in harm's way in the service of their country. Their narratives are offered in an effort to assist others to be better prepared when they are called out on operations. In addition, the lessons that leap from the pages should serve to generate discussion and debate in order to determine better practices and policies that will serve to make the CF more effective, efficient and operationally ready.

INTRODUCTION

CRUCIBLE OF SUCCESS: THE SENIOR NON-COMMISSIONED OFFICER CORPS

Bernd Horn

> *Non-commissioned officers [NCOs], promoted from the ranks, enforced discipline, managed routine administration, and socialized recruits – and sometimes their commissioned superiors – in military ways. By common consent, they formed the backbone of their unit.*[1]
>
> Canadian Military Historian, Desmond Morton

Desmond Morton's description of the senior NCO in the First World War seems timeless. One could argue that his description was as accurate in the early twentieth century as it is today. Few would argue that one of the fundamental hallmarks of a modern military is the existence of a professional senior non-commissioned officer (NCO) corps. After all, it is the senior NCOs that are instrumental in ensuring that the day to day affairs of the military are executed – whether for training, discipline or the conduct of operations. They concentrate on force preparation, specifically training individuals and small teams to ensure that they are motivated, well-trained (i.e. meet the standards of performance for a given trade, skill or function) and ready to carry out their mission. They are also instrumental in the professional development of young soldiers and junior NCOs. On operations, they ensure those attributes and skills learned in training are put into practice. In short, senior NCOs get the day to day jobs of the military done.

Although senior NCOs have always been the backbone of any army, their importance has grown in the modern battlespace. Operations in the chaotic, complex security environment of today necessitate more than ever the presence of a strong, professional senior NCO corps. The safe, templated and well-known Cold War paradigm disappeared almost overnight. The new security environment that filled the vacuum was marked by complexity, ambiguity, and an ever present media, as well as nefarious enemies and threats that were embedded in the context of failed and failing states. This reality overloaded a traditional, conservative and intellectually inflexible military institution that until this point saw the world in terms of absolutes.

In the post-Cold War era, the sanctity and security of the old conventional wisdom that held true for almost half a century was shattered. The carefully prepared plans for the defence of Germany became irrelevant. Gone was the familiar and carefully templated Soviet enemy.[2] Military leaders, both officers and senior NCOs, whose entire careers were rooted in, and defined by, the comfortable predictability of the Cold War were faced by a brave new world. Conflict had become exponentially more complex and unpredictable. "Today's world is without precedent," cautioned French military analyst Phillippe Delmas, "it is as different from the Cold War as it is from the Middle Ages so the past offers no basis for comparison."[3]

As such, the transition from Cold War, to Post Cold War to the Post 9/11 era greatly impacted the senior NCO Corps. During the Cold War, senior NCOs were an integral element in the army that emphasized large mechanized formations that would fight an attritional symmetrical battle on the central European front. Although integral to the day to day training and discipline of the military, operationally they were but a microcosm of a bigger whole. Battle was waged in large mechanized formations, supported by distant weapon systems, air armadas and fleets at sea.

Then suddenly things changed. In the post-Cold War era, the allure of a peace "dividend" and a promise of a revolution in military affairs generated beliefs that the necessity of large ground forces would wane. Moreover, the trend towards a more technologically reliant and functional military institution fueled talk of possibly down-sizing, re-roling and transforming the senior NCO corps. The new era of peace support operations now occurred in such volatile areas as the Former Yugoslavia, Somalia and Haiti, where the world grappled with chaos and violence of failed and failing states, thereby re-emphasizing the role of small unit leadership at the sub-unit and platoon levels.

More cataclysmic yet was the aftermath of the terrorist attack on the Twin Towers of the World Trade Centre on 9 September 2001 (9/11). As the

Americans launched their global war on terror, enemies of the US and their allies adopted asymmetric means to combat the militarily and technologically superior military forces arrayed against them. Doctrinally, an asymmetric threat is a concept "used to describe attempts to circumvent or undermine an opponent's strengths while exploiting his weaknesses, using methods that differ significantly from the opponent's usual mode of operations." [4]

In short, the enemy adopted fourth generation warfare (4GW), which refers to a nonlinear, asymmetric approach to war in which agility, decentralization and initiative are instrumental to success. 4GW is nonlinear, widely dispersed and undefined. The distinction between "civilian" and "military" also often disappears. As witnessed in Iraq and Afghanistan, the militant forces combating allied coalition forces utilize every advantage possible, such as targeting civilians; using religious or medical sites as sanctuary, staging areas, logistical centres or actual locations to launch attacks; as well as improvised explosive devices (IEDs) and suicide bombers.

Herein lies the difficulty for the practitioner. Military leaders are required to operate in, and be comfortable with, ambiguous and uncertain surroundings. Their options for the type of, if not the use of, force is often restricted. In addition, born of necessity, they require the capability of adapting physically and theoretically to changes not only in their immediate operational area but also in the larger international security environment. The new defence environment also demands that individuals, units and formations be agile, flexible and capable of responding to the unforeseen and unexpected. It becomes a constant battle – the person on the ground interfacing with the population – making decisions, often life or death decisions, on a constant daily basis. It becomes, once again, a senior NCO's war.

It is for this reason that the current security situation necessitates that all military institutions have a professional senior NCO Corps. After all, they represent the cerebral cortex of a unit and have always been, and will continue to be, essential to the effectiveness of the army, specifically because of their traditional roles and responsibilities. The current struggles in both Iraq and Afghanistan demonstrate that regardless of the ascendancy of technology, the senior NCO Corps will always be required to satisfy a function that is indispensable to the viability of a military institution. Technology changes the tools with which armies fight, but it does not change the reality of man's role in conflict. War will never be clean or sterile. Clinical strikes by stand-off, long range precision guided missiles, designed to limit collateral damage and minimize friendly casualties, are only effective if the damage wrought is of significance to the recipient. As US Marine Corps Lieutenant-General Paul van Riper has postulated, what if the enemy simply ignores the attack?[5]

Although technological and scientific breakthroughs cannot be ignored, the technological edge that is potentially provided can be of limited utility in situations such as are presently being experienced in South East Asia and those that were experienced in Rwanda, Somalia and Haiti. "The warrior's trade," argues George Friedman, chairman of Strategic Forecasting in Baton Rouge, "will remain one of courage, dedication and suffering." He adds, "Precision-guided munitions will not render war antiseptic, any more than did the tank or crossbow or bronze armor. Technology changes how men fight and die, but it does not change the horror and glory of battle, nor does it change the reality of death."[6] Similarly, Dr. Williamson Murray, a retired army officer commented, "What makes this techno-craze so dangerous is that it flies in the face of 2,500 years of history, not to mention modern science. Friction, ambiguity, chance and uncertainty are not merely manifestations of inadequate communications and technology that U.S. military organizations in the next century may overcome, but rather manifestations of the fundamental nature of the world, where if something can go wrong, it will."[7] Van Riper argues that "Real war is an inherently uncertain enterprise in which chance, friction and the limitations of the human mind under stress profoundly limit our ability to predict outcomes; in which defeat to have any meaning must be inflicted above all in the minds of the defeated..."[8] Quite simply, technology can assist and enhance the capabilities of fighting soldiers, but in the end it will never be able to totally replace them. War will always remain a truly human endeavour.

It is within this context, namely the ugly reality of conflict, complete with its ambiguity, fear, friction and uncertainty that the prodigious importance of the senior NCO emerges. In spite of the changing natures of conflict, society and technology, there are a number of core responsibilities of the senior NCO corps that cannot change without imperiling the effectiveness of a nation's military. This is plainly visible when the actual functions of the senior NCO are examined. First, he provides the critical link between the soldiers and the officer corps and vice versa. In this vein, the senior NCO often fulfills a role similar to that of an ombudsman for the rank and file. Furthermore, he is an administrator, trainer, mentor and at times parent to those entrusted to his charge. In addition, he is also a motivator and disciplinarian, as well as a combat leader and tactician.

It is clearly evident from this superficial summary that the senior NCO corps is the moral cement – the glue – that holds an army together. "The choice of non-commissioned officers," insisted Baron Frederick von Steuben, "is an object of the greatest importance." He explained, "The order and discipline of a regiment depends so much upon their behaviour, that too much care cannot be taken in preferring none to that trust but those who by their merit and good conduct are entitled to it." He added, "Honesty, sobriety, and a remarkable attention to every point of duty, with a neatness in their dress, are indispensable requisites."[9] This reality changed little in a century and a half. In 1942, General Bernard Law Montgomery echoed Steuben's sentiments when he described non-commissioned officers as the backbone of any unit.[10]

Needless to say, Steuben and Montgomery were neither the first nor the last commanders to make such a resolute assertion in regards to the senior NCO cadre. The reason for these accolades merits closer scrutiny. After all, to fully understand the importance of the senior NCO Corps, one must have a full and comprehensive understanding of their traditional roles and responsibilities.

Land Force Command, in its seminal doctrinal publication *Canada's Army*, states that senior NCOs "...are the link connecting soldiers to their officers and officers to their soldiers. Their role is to translate the intentions of commanders into action."[11] This interpretation is rooted in law. The *Queen's Regulations and Orders* specify the general responsibilities of all non-commissioned officers. These include the observance and enforcement of various regulations, rules, orders and instructions; promoting good discipline, welfare and efficiency of all who are subordinate to the member; and reporting to the proper authority any infringement of the pertinent statutes, regulations, rules, orders and instructions governing the conduct of any person to the Code of Service Discipline.[12]

Simply put, the senior NCO conducts the daily business of an army. He is expected to carry out instructions to achieve the institutional aim, as well as ensure the well-being of the soldiers. He is also trusted to pass on the ethos and traditions of the military in general and the respective regiment in particular. Furthermore, he is required to pass on and explain directives from his superiors to his subordinates, as well as to ensure that all direction is efficiently and effectively executed.

In addition, there is a justified expectation by officers that their senior NCO has a more up to date, as well as more accurate, insight into the general feeling and morale of the rank and file and that potential problems or discontent are passed up the chain of command. This sentiment was clearly articulated in 1880, by then Colonel William Otter, in his *Guide: A Manual for the Canadian Militia.* Otter explained that it was essential that NCOs know intimately all of the characters and capabilities of their men. In regard to the Sergeant-Major, he insisted that he must be an advisor to the unit Adjutant and in general must act as "the eyes, ears and conscience of the battalion." For this reason, Otter stated, that a sergeant-major had to be a "man of sound sense and judgment, of good temper and kind but firm in execution of his duty."[13]

It is in this knowledge of, and closeness to, the men that the senior NCO proves his greatest strength. For only with a complete and proper understanding of an individual can you truly elicit their greatest potential. Moreover, the NCOs' full comprehension of their subordinates, as well as their pivotal role as the link between the soldiers and the officers, positions them to act as an ombudsman for their men. Discontent and morale problems can often be avoided or quickly resolved by timely intervention or advice. In this vein, senior NCOs provide clarity and context to transgressions in behaviour and perceived affronts or injustices. In addition, they provide voice for aggrieved, intimidated and/or over-anxious subordinates, particularly young soldiers.[14] In essence, their maturity, experience and knowledge provide a steadying influence in an institution that is steeped in human endeavour with all the confusion and frailty that inherently accompanies such enterprise.

Notwithstanding the importance of the aforementioned roles, instrumental to the vitality of any army is the quality and capability of its fighting forces. Here lies a fundamental and all important role of the senior NCO – that of the trainer. Training is of immense importance in influencing behaviour. It is also critical in imparting those abilities and skills that will determine the survival of individuals in combat and the success of an army on the battlefield. Although seldom articulated in this manner, the performance of an army rests largely on the skill and professionalism of its senior NCO corps. "The NCO is expected to provide advice on everything," stated Chief Warrant Officer (CWO) Randy Northrup. He asserted, "NCOs share knowledge experience, honesty, loyalty and support."[15] Similarly, "it is the NCO," explained Command Sergeant-Major Wilbur V. Adams, Jr., "who directs individual training, builds physical readiness, and teaches core values of leadership and moral courage."[16]

Young recruits are introduced and taught the basics of soldiering and military life. Upon joining their unit, soldiers continue to learn basic and advanced skills necessary to fulfill their role as combat troops. The proficiency of the individuals creates the foundation upon which collective training can be conducted. "We had exceptionally good sergeants," recalled Corporal Denis Flynn, of the 1st Canadian Parachute Battalion in World War II. "They knew how to organize and prepare soldiers," he explained, "That was their key function – teaching soldiers how to do their job and survive."[17] His simple recollection underlines a salient point. If the senior NCO corps is deficient, the foundation of a unit, and the army in general, will be weak and, thus, condemn the entire structure to collapse. It is not surprising that Colonel Otter believed that the fundamental success of a unit was "in a great measure dependent on the alertness and skill of the NCOs."[18] He believed that that NCOs "are the very 'stuff' of the army."[19] CWO Greg Lacroix, the Canadian Forces (CF) CWO stated, "I have been saying for some time now that these operations [war in Afghanistan] are riding on the backs of the NCOs. Unfortunately they come home and we burden them with more demands for training our men and women."[20]

What must be remembered is that the NCOs' skill to train unit personnel is not limited exclusively to soldiers. It is commonly recognized as an age old truth by military personnel worldwide that there is nothing more dangerous than a second-lieutenant. This threat is mitigated to a large extent by the senior NCO Corps. Notwithstanding the fact that a young officer of this rank is in a superior position of authority to senior NCOs, in reality their education and training, in regards to unit and military life, is actually shaped to a great degree by their NCO subordinates. *Canada's Army* institutionalizes this concept. It articulates that senior NCOs "... have an important responsibility in teaching newly joined officers hands on skills in the mechanics of soldiering and leadership. This includes offering advice, helping solve problems, and providing feedback and information."[21] Similarly, American First Sergeant Jeffrey J. Mellinger in an open letter to his NCO Corps explained, "Training your platoon leader is not only your job but your responsibility. If he fails, the platoon fails, and so do you." He added, "As the senior and most experienced NCO [platoon sergeant] in the platoon, you must pass on the benefit of that wisdom and experience to your platoon leader as well as to the soldiers."[22]

The importance of the NCO's role as a trainer was clearly recognized by the German military. For example, in the inter-war years, NCOs were career professionals who were carefully selected and enjoyed considerable prestige in society. Upon completing his military career, the German NCO was guaranteed a pension of 1,500 marks, as well as preferential hiring in the civil service, railroads, and the postal system.[23] Tom Clancy, a recognized expert on military affairs, and American General Fredrick M. Franks, a former commander of Training and Doctrine Command (TRADOC) and the Commander of VII Corps during the Gulf War in 1990-91, provide a more contemporary view of the importance of the senior NCO corps as trainers. They wrote:

> *The years of fighting in Vietnam had drawn Europe-based forces down to unacceptable strengths. Worse, the insatiable appetite for personnel had stripped our forces of officer leadership, and almost destroyed the Army's professional non-commissioned corps, long the backbone of the Army. A series of hasty training programs to fill depleted ranks had left the Army with NCOs who all too often were poorly trained in basic leadership techniques. Because the NCO is the first-line leader in the Army, the one person primarily responsible for the basic individual soldier skills on which every successful operation depends, training and discipline suffered. In some cases, it went to hell.*[24]

Lockstep with the requirement to train individuals is the imposition and enforcement of discipline. Once again, the senior NCO has always been the linchpin. As already stated, universally, the NCO is the crucible of success in any army. He inculcates discipline through personal example and training. By his words and actions he defines to the soldiers what is acceptable behaviour and what is not. It is the senior NCO that often most effectively promulgates, explains and enforces directives, orders and policies. This should not be surprising. This is the result of the close, direct and daily contact between a senior NCO and his men. "From the perspective of a line company enlisted man," explained one World War II veteran, "our day to day lives were impacted intensely by NCOs. They were the flesh and blood leaders we knew."[25] This is why historically senior NCOs have been an invaluable resource to the army and often the difference between success and failure on operations.

An example of the catastrophic results of a failure in discipline was clearly evident in the Canadian Airborne Regiment during its deployment to Somalia in 1992-93. Years of poor manning practices, a large number of weak NCOs (as well as officers) and a failure to maintain discipline within the unit, led directly to the torture killing of a Somali youth. Incredibly, although evidence has shown that both NCOs and officers were aware, and/or should have been aware that a beating of a detainee was transpiring, nothing was done to stop the killing in time. This transgression became cataclysmic, and combined with other events, eventually led to the disbandment of the unit; the first in Canadian history to be done so as a result of perceived disciplinary problems.[26]

The importance of discipline in the military to ensure the subjugation of personal will to that of the institution, as well as the NCO's role in achieving this, is unquestioned. But the true value of the senior NCO cannot be really appreciated by simply detailing individual responsibilities. This is to miss their historical contribution to the army. In totality, beyond administrator and trainer, the senior NCO is a mentor, motivator and often parent figure to his soldiers. His actions instill trust, loyalty and ensure performance in the most trying conditions. During the Second World War, "mother" became German slang among the field soldiers to denote the senior sergeant in a company. The rationale behind this is not difficult to comprehend. For instance, Karl Fuchs, who served in such a position, wrote home to his father during the war, "I've become such an integral part of my company that I couldn't leave it ever again."[27] Similarly, Hans Werner Woltersdorf asserted, "my unit was my home, my family, which I had to protect."[28]

Dr. John McManus's study of the American combat soldier in World War II replicated the German experience. He found that in most cases the natural leaders in combat were the NCOs. One veteran's account was typical of his finding:

> *...[The platoon sergeant] was caring and wanted to make sure his men had what they needed. He boosted morale. He checked weapons every day, made sure the men had dry socks. He hovered over us like a mother hen.* [29]

As already alluded to in the anecdotes, nowhere is the senior NCO's importance easier to ascertain than during combat or operations in general. It is here that the senior NCO proves his undisputed title as the backbone of the army. He becomes the key to operational success, turning commander's intent into action. It is in operations that the sum of all responsibilities and tasks of the senior NCO are welded together – where he acts as technician, tactician and leader.

The senior NCO's role as technician is clearly evident. Their comprehensive knowledge of weapons and equipment as a result of their technical training and the instruction of others make them an invaluable advisor to officers and instructor/supervisor to soldiers. Their ability to ensure the care, maintenance and first line repair of various unit material is instrumental to unit effectiveness, particularly during the immediacy and stresses of operations.

But of paramount importance to the success of a unit is the senior NCO's role as tactician and combat leader. Many may argue the sobriquet of tactician. After all, he simply carries out orders. During World War I the German NCO, similar to that of most other nations, "primarily served his men as a model of military toughness, a disciplinarian and enforcer of military regulations and 'discourager' to the 'Drueckeberger (slacker) in combat, but not as a model battlefield tactician."[30] But this soon changed with the evolution of mechanized warfare. In World War II, the concept of the senior NCO as a tactician was simply unavoidable. If for no other reason, casualties forced them to take on this responsibility. "As a Platoon Sergeant, 4 Platoon, 'B' Company, [1st Canadian Parachute Battalion]," recalled Sergeant R.F. Anderson, "there were many times in the field when I had no Officer by reason of casualty or illness. This meant a delay in getting a replacement which had to come from England, so in many cases companies were led by senior NCOs..."[31] Fellow paratrooper Corporal Dan Hartigan agreed. "We lost over 50 percent of our officers on D-Day," he explained, "15 of the 27 I believe."[32] He added, "the fighting in the weeks that followed turned from an officer's war to a senior NCO's war." McManus's study supports this anecdotal evidence. "With surprising frequency," he concluded, "combat soldiers looked to their sergeants for leadership day in and day out in combat. A major reason for that was the turnover in officers."[33]

But the claim of tactician is not built exclusively on the senior NCO's ability to become a battlefield replacement for officer casualties. The nature of his position within the army, not to mention the established doctrine, necessitates this. One of Israeli's outstanding field commanders, Yigal Allon, proclaimed that the great battles of the 1948-49 War, the Sinai Campaign in 1956, and the Six Day War in 1967, were "won in the NCO's courses of the *Haganah* and the *Palmach*." He explained:

> *The most brilliant plan devised by the most capable general depends for its tactical execution on the section-leaders. Poor section-leaders may ruin the best-laid plans; first-rate section-leaders will often save badly devised plans. This for one simple reason: the section-leader is the sole level of command that maintains constant and direct contact with the men who bear the brunt of the actual fighting. It follows, then, that the section-leader is to be trained as a tactical commander and as an educator of his men. [In the Israeli Defence Force] section-leaders are trained to command independently in the field in every instance in which they are required to operate alone with their units. In regular combat, moreover, when the section-leader acts within the framework of his platoon and under orders from his superior officer, he still requires a high standard of knowledge and an ability to sum up the situation.*[34]

In the current ambiguous, complex and volatile security environment, the senior NCO plays a vital role. In an environment where winning the hearts and minds of the local population is in direct competition with an enemy who will exploit any advantage and any perceived weakness the example, guidance and direction of the senior NCO on the ground represents the difference between life and death, for both civilians and military members, on a daily basis.

Whether the designation of tactician is accepted or not, in the end, the NCO's primary function in combat is still one of leading his men against an enemy as far as possible undiminished in strength to achieve a military objective. To accomplish this, he must provide leadership, direction and guidance. Moreover, he must render psychological, as well as physical, support by furnishing an example of courage, as well as a sense of calm and a presence of mind regardless of surrounding turmoil. Additionally, throughout, the senior NCO must remain with the element that is closest to the enemy and, thus, danger.

This reality is the ultimate proof of the NCO's undeniable importance to the efficiency of the army. It is generally accepted that leaderless groups usually become inactive. The provision of this leadership, often in the face of chaos, routinely falls to the senior NCO. This is only natural, after all, in relation to their men they represent the first level of command. To the NCO cascades the easily quoted, yet more difficult to execute, task of "closing with and engaging the enemy."

In this endeavour, the NCO's presence becomes all important. Of a group of combat veterans surveyed, 89 per cent emphasized the importance of getting frequent instructions from leaders when in a tight spot. They felt that coolness in combat was contagious. In fact, 94 per cent of the respondents believed that "men feel they fought better after observing other men behaving calmly in a dangerous situation."[35] Quite simply, "men like to follow an experienced man....[an] experienced man knows how to accomplish objectives with a minimum of risk. He sets an example of coolness and efficiency which impels similar behaviour in others."[36] For this reason it is not surprising that studies of World War II combat veterans indicated that in respect to courage, 42 per cent expected it of their NCOs, who they most closely related to and relied on, while only 30 per cent expected it of their officers.[37]

Anecdotal evidence provides some graphic examples. "I was getting jittery," confessed Private Alexandre Huton, "but the sergeant was steady as a rock. What a soldier!" He further elaborated:

> *We moved again, this time over a cratered field, the sergeant always moving in front of us. Twice he stopped and dug his toes in the dirt, then he lead [sic] us around a mine. That was typical of the sergeant, he tried everything himself first before he would allow the rest to follow. When we came to wide dikes the sergeant swam across first to see if we could make it across. We moved up on the made road again, the sergeant stopping to cut the Jerry telephone wires with his bayonet and again we had to fall flat while Jerry field guns rolled past us. The sarge was always taking note of their equipment and direction. He*

didn't miss a trick...I'd lost track of where we were going but the sarge was leading us and we took his word for it.[38]

Similarly, the actions of another NCO, Company Sergeant-Major (CSM) John Kemp, reinforce the above image. One official report recorded, "CSM Kemp who had assisted in organizing the men for the attack, with complete disregard for his own safety, led his men against the farm house in spite of the heavy fire. By his personal example, he enabled the small force to overrun the enemy defences and capture the position."[39] In another example, Private Bill Gates of the Royal Marines recalled, "I rushed on behind one of our Sergeants, a great big bloke who seemed afraid of nothing, and I felt so long as I stayed behind him I'd be OK."[40]

In a similar vein, assessments of combat soldiers provide a clear idea of not only the expectations of the men but also the importance of the NCO. For example, CSM Charlie Martin of the Queen's Own Rifles was described by his men as a Riflemen's dream sergeant. He was by all accounts "an outstanding soldier," who could always be found up front "even if there was only a small patrol action." To those who knew him, he "was a 'come on' not a 'go on' leader."[41]

Indisputably, the senior NCO's responsibilities as administrator, trainer, disciplinarian, mentor, leader and tactician are pivotal to the efficacy of an army. Their example, close proximity and relationship to their soldiers enable them to motivate individuals in the face of fear and danger and accomplish tasks. Studies from Korea and Israel reveal that unit NCOs provide the emotional support for the squad and manage this by a mixture of modeling, sustaining and teaching.[42] Quite simply, they are the glue that holds an army together. Thus, regardless of technological and scientific developments, as long as human conflict exists, the requirement for men and women to go in harm's way will remain. And, so too, will the traditional roles and responsibilities of the senior NCO.

That being said, there are of course some things that must evolve. The new complexity in conflict discussed earlier,[43] combined with the CNN effect, often couched in terms of the "strategic soldier," and exacerbated by societal expectations, necessitate that a greater emphasis be placed on the education of the NCO instead of the long-standing exclusive focus on training.

"The large formation, closely controlled and highly supervised troops in the warfare models of the cold war era have given way to scattered small units in distant countries," argues Colonel Paul Maillet, the Department of National Defence (DND) director of Defence Ethics, "who have reduced support readily available in ambiguous and high intensity ethical situations." He adds, "A wrong decision in the glare of the media can have far reaching consequences that can affect peacekeeping mandates and strategic and national policies and aims."[44]

Herein lies the paradox: this realization underlines the need for both the retention of the traditional roles and responsibilities of the senior NCO corps, as well as the requirement to change the manner in which the individual NCO is prepared to fulfill his function. The traditional stress on training, that is "a predictable response to a predictable situation," must be better balanced with education, defined by Professor Ron Haycock as "the reasoned response to an unpredictable situation – critical thinking in the face of the unknown."[45] Simply put, the prescribed application of ideas and methods, as well as drills and checklists, have a purpose and functional utility, but this methodology is no longer enough to equip leaders to cope with and function in the complex post modern world. "There are no standard 'drills'," remarked Art Eggleton, the Minister of National Defence (MND), "for the many complex challenges that our troops confront in places like Bosnia."[46] This reality has become even more paramount in Afghanistan.

Undeniably, senior NCOs, in the same manner as officers, must be taught how to think and use abstract concepts to assist in the resolution of practical problems that they may face. They must expand their knowledge and acquire a broader outlook, as well as develop greater socio-political skills. Furthermore, they must become comfortable with ambiguity and change. Critical thinking and innovation must become their guiding light instead of the traditional heavy reliance on written procedures given in technical publications and uni-dimensional experience. To achieve this end, education must be aggressively pursued.

The requirement for greater education is also rooted in the necessity of dealing with today's soldiers. Pierre de Reil, French Minister of War in 1793, wrote "as long as the soldier believes himself equal in intelligence and knowledge to his commanders, he will not obey."[47] With the average recruit entering the CF increasingly with a minimum of a high school education or better, the senior NCO is required to continually advance his own base of knowledge so that he can not only deal with the ambiguity and complexity of operations, but also so that he can teach, train, motivate and lead progressively more sophisticated subordinates.[48] The old days of drill sergeants castigating recruits or soldiers with derogatory, expletive filled expressions, or taking a "problem individual" out behind the woodshed to "sort him out" are not only inappropriate but are also no longer effective. Senior NCOs must understand and be capable of motivating their new charges, despite generational differences and new societal expectations, norms and values. Once again, education is the key.

Increasing the education of the senior NCO corps is linked to another fundamental change that must occur – employment of individuals by abilities and skills. With the evolution of military affairs and the increasing educational profile of both NCOs and officers, the traditional practice of filling organizational positions exclusively by rank must evolve into a more flexible system of focusing on individual capability. For instance, staff positions in various level headquarters or schools that are normally filled by junior officers, United Nations (UN) staff positions or military observers, or even some operational positions such as platoon commander billets, can be filled as a normal posting (and not just as a temporary expedient) by qualified senior NCOs. This dramatic change is necessary for a number of reasons. First, it helps alleviate organizational shortfalls that stem from the current CF recruiting and retention crisis. Second, it fulfills the requirement to recognize individuals for their abilities and rewards those who have put forth the effort to increase their education and skill sets. Third, it allows for personal and professional challenge and growth, so necessary for attracting and retaining individuals in the CF.[49]

The challenge before the CF is a daunting one – how to retain the timeless strength of the senior NCO corps that has made it the crucible to success of the army, yet ensure its evolution to keep it a viable and relevant force in the future. The key to success lays in understanding the role and responsibilities of the senior NCO corps and ensuring that they are given the educational foundation to provide them with the necessary knowledge, as well as thinking strategies, to deal with ambiguity and change. In addition, it is also important to recognize that as long as there is human conflict, regardless of technological and scientific breakthroughs, their ability to train and lead well-trained soldiers into chaos, danger and turmoil will remain the definitive test of an army's efficacy.

ENDNOTES

1 Desmond Morton, *When Your Number's Up* (Toronto: Random House of Canada, 1993), 95.

2 In 1987, Georgi Arbatov, advisor to Mikhail Gorbachev, announced to the West, "We are going to do to you the worst thing imaginable – we are going to deprive you of your enemy." Pascal Boniface, *The Will to Powerlessness. Reflections on Our Global Age* (Kingston: Queen's Quarterly, 1999), 37.

3 Phillippe Delmas, The Rosy Future of War (New York: Free Press, 1995), 213.

4 Colonel W.J. Fulton, DNBCD, "Capabilities Required of DND, Asymmetric Threats and Weapons of Mass Destruction," Fourth Draft, 18 March 01, 2/22.

5 Paul van Riper and Robert H. Scales, "Preparing for War in the 21st Century, *Parameters*, Autumn 1997, 11. See also Charles J. Dunlap, "21st Century Land Warfare: Four Dangerous Myths," *Parameters*, Autumn 1997, 27-37. The four myths are: 1- Our most likely future adversaries will be like us. 2- We can safely downsize our military in favour of smaller, highly trained forces equipped with high-technology weapons. 3- We can achieve information superiority and even dominance in future conflicts. 4- Modern technology will make future war more humane if not bloodless.

6 George and Meredith Friedman, *The Future of War* (New York: St Martin's Griffin, 1998), xi.

7 Williamson Murray, "Does Military Culture Matter?" *Orbis*, Winter 1999, 37-38.

8 Van Riper and Scales, 5.

9 Major-General M.K. Jeffery, "The Non-Commissioned Officer of the Future Army: Introduction," *Backbone of the Army. Non-Commissioned Officers in the Future Army*, ed. Douglas L. Bland (Kingston: McGill-Queen's University Press, 2001), 2.

10 John A. English, *Failure in High Command* (Ottawa: The Golden Dog Press, 1995), 318.

11 Land Force Command, *Canada's Army* (Ottawa: DND, 1998), 52.

12 *Queen's Orders and Regulations*, Article 5.01.

13 Major-General Sir William Dillon Otter, *The Guide: A Manual for the Canadian Militia, 9th ed.* (Toronto: Copp, Clark Company, 1914), 21-22. See also Ronald G.

Haycock, "The Stuff of Armies: The NCO Throughout History," in *Backbone of the Army. Non-Commissioned Officers in the Future Army*, ed. Douglas L. Bland (Kingston: McGill-Queen's University Press, 2001), 9-23.

14 An interesting anecdote provides some clarity. During the WWII a promising young sergeant was asked to take his commission. He refused. "You see," he said, "what I am really interested in is the personal contacts with the men and I know that if I become an officer I should not have the same chance of helping them as I have now." Lieutenant R. Bernays, "Man-Officer Relationships," *The Army Quarterly*, Vol XLVI, No. 2, August 1943, 253.

15 Chief Warrant Officer Randy Northrup, interview with author, 22 January 2007.

16 Wilbur V. Adams, Jr., "The Non-Commissioned Officer in the United States Army: Leading by Example," *Backbone of the Army. Non-Commissioned Officers in the Future Army*, ed. Douglas L. Bland (Kingston: McGill-Queen's University Press, 2001), 87.

17 Corporal Denis Flynn, interview with author, 18 April 2001.

18 Colonel Otter, as quoted in Haycock, 18.

19 Haycock, 19.

20 Chief Warrant Officer (CWO) Greg Lacroix, Canadian Forces CWO. E-mail to author 10 September 2007.

21 *Canada's Army*, 52-53.

22 First Sergeant Jeffrey J. Mellinger, "Open Letter to Three NCOs," *Infantry*, May-June 1989, 20. This aspect of a Senior NCO's responsibility is universal. The famous British Lieutenant-Colonel Colin Michel (of the Argylls) wrote, "My platoon sergeant, a tough little man called Dempsey was obviously just the professional I needed to nurse me into my new job." Colin Mitchell, *Having Been a Soldier* (London: Mayflower Books, 1969), 41.

23 James S. Corum, *The Roots of Blitzkrieg* (Kansas: University of Kansas, 1992), 11.

24 Tom Clancy with General Fred Franks, Jr. (ret.), *Into the Storm* (New York: Berkley Books, 1998), 85.

25 John C. McManus, *The deadly brotherhood. The American combat soldier in World War II* (Novato: Presidio, 1998), 219.

26 See Report of the Commission of Inquiry into the Deployment of Canadian Forces to Somalia, *Dishonoured Legacy. The Lessons of the Somalia Affair* (Ottawa: Canadian

Government Publishing, 1997), Vol. 1, 244-247 & 324; and Vol 2, 429-470. See also Bernd Horn, *Bastard Sons - A Critical Examination of the Canadian Airborne Experience, 1942-1995* (St. Catharines: Vanwell Publishing, 2001).

27 Stephen G. Fritz, *Frontsoldaten. The German Soldier in World War II* (Lexington: University of Kentucky Press, 1995), 18-19.

28 Ibid., 19.

29 McManus, 218-220.

30 Karl H. Theile, *Beyond Monsters and Clowns. The Combat SS* (New York: University Press of America, 1997), 99.

31 Sergeant R.F. Anderson, letter to author, 19 December 2000.

32 Corporal Dan Hartigan, interview with author, 30 October 2000.

33 McManus, 202.

34 John A. English, *On Infantry* (New York: Praeger, 1984), 191.

35 John Dollard, *Fear in Battle* (Westport, Connecticut: Greenwood Press, Publishers, 1944), 28. See also Elmar Dinter, *Hero or Coward* (London: Frank Cass, 1985), and S.J. Rachman, *Fear and Courage* (San Francisco: W.H. Freeman and Company, 1978.

36 Dollard, 44. It is generally accepted that "the presence of careful and thoughtful leadership builds up a force which helps resist fear."

37 Elmar Dinter, *Hero or Coward* (London: Frank Cass, 1985), 53.

38 L.S.B. Shapiro, "I Dropped Alone," *MacLean's*, 1 August 1944, 5-6.

39 Jean E. Portugal, *We Were There. The Army. A Record for Canada, Vol 2 of 7* (Toronto: The Royal Canadian Military Institute, 1998), 968.

40 Edmund Blandford, *Two Sides of the Beach. The Invasion and Defense of Europe in 1944* (Edison, NJ: Castle Books, 2001), 222.

41 Ibid., 706.

42 Colonel Ian Palmer, "The Emotion That Dare Not Speak Its Name?" The British Army Review, Summer 2003, No. 132, 36.

43 US Army generals who have served in both classic warfare and conflict resolution environments assert that the type of complexity found in conflict resolution, such as Bosnia, "does not exist on the modern battlefield." They add that "Bosnia holds a level of complexity and frustration that is new." Quoted in "The Civilian-Military Gap," Online NewsHour, www.pbs.org/plweb-cgi/fastweb/getd, accessed 10 November 1999.

44 Colonel J.P.M. Maillet, "Defence Ethics Program Ethics and Operations Project," memorandum dated 20 January 2000.

45 Dr. Ronald Haycock, former Dean of Arts, Royal Military College (RMC), "Clio and Mars in Canada: The Need for Military Education," presentation to the Canadian Club, Kingston, Ontario, 11 November 1999.

46 Speaking Notes for the Honourable Art Eggleton, Minister of National Defence, Canadian Forces College, 19 June 2000, 10.

47 John A. Lynn, *Bayonets of the Republic* (Chicago: University of Illinois Press, 1984), 89.

48 More and more one hears of the "corporal with a Masters Degree" as the way of the future. Although there are several cases of this in the CF, largely in the Reserves, a few does not a trend make. The argument that soon there will be no marked difference between the NCO cadre and the officer corps may be considerably premature. These apocalyptic warnings are not new. "The results of popular education have borne fruit," wrote Major R.A.C. Radcliffe in 1943, "and there are to-day a considerable number of men in the ranks who have learnt to think and reason for themselves." He went on to state that the improved education of the men in the ranks made them "more critical of their officers in every way." Major R.A.C. Radcliffe, "Officer-Man Relationships," *The Army Quarterly*, Vol. XLVI, No. 1, May 1943, 114-116. In a similar vein, another account from 1943 revealed, "In the old Army there was a blind obedience, the result of strict discipline, which was often confused with loyalty to one's officer's and to the Army. To-day there is a much more questioning quality ... the NCOs and men nowadays are not blindly loyal to their senior officers just because the latter happen to hold the King's commission. They first wish to satisfy themselves that their officers are thoroughly capable, and, as intelligent men, they take mental note of every action, look and word of their seniors...." C.W. Valentine, "Army Morale and its Relation to Discipline and Efficiency," *The Fighting Forces*, Vol. XX, No. 1, April 1943, 24. Although the need for education will become increasingly important at all levels in DND and the CF, it will remain as the predominant difference between non-commissioned members and the officer corps.

49 See Canada, *People in Defence Beyond 2000* (Ottawa: DND, 2001).

CHAPTER 1

BUILDING FOR THE FUTURE: THE CONSTRUCTION OF CAMP MIRAGE, OPERATION APOLLO, ROTATION ZERO

Charles Malchow

Following the terrorist attacks on US soil of 11 September 2001 and the subsequent commitment by the Canadian government to deploy troops to Afghanistan, it became quickly apparent that a forward operating base was required near the site of the conflict. The following is an account of how Camp Mirage was built, including many of the successes and challenges that were faced.

By mid-morning, Thursday 20 December 2001, our Second in Command (2IC), a captain, received a call instructing our unit to return immediately to Moncton, New Brunswick. The message left no room for interpretation: "overseas deployment is forthcoming. RTU [return to unit] soonest."

At the time that the message arrived, we were completing a recce/survey for Canadian Forces Base (CFB) Trenton's operations and training cell. We are construction engineers. We are members of 3 Specialist Engineering Team (SET), 1 Construction Engineering Unit (CEU) Moncton, a Deputy Chief of Defence Staff (DCDS) asset, ready to deploy on short notice anywhere in the world. Our unit has three SETs, manned and equipped to be "First In and Last Out."

By late Friday afternoon, 21 December, already back in Moncton, there was a buzz of activity in the Unit. Our SET in command (i/c), a captain, scheduled an orders group (o-group) with all members to pass on what limited information there was. This meeting was slated for 1800hrs, Saturday, 22 December, in our main conference room. Every day is a workday.

"Gents we are leaving on Boxing Day for Greenwood, Nova Scotia. Our mission is the design of a bed-down camp in support of Operation (Op) Apollo. Our destination is a country in South West Asia," our SET IC informed us. We were advised that a Site Activation Team (SAT) had already completed its mission overseas and a site had been found with all the utilities required for our design needs. The Site Activation Officer would be joining us in Greenwood. We also received our Temporary Attached Posting Message which indicated that we would be a component of the Long Range Patrol Task Force (LRPTF) from CFB Greenwood, arriving in theatre on 29 December 2001.

By 26 December, Boxing Day, the team was excited and ready to roll. Our deployable equipment had been checked and loaded onto a 5-ton truck just prior to lunch. Our Commanding Officer (CO), Regimental Sergeant-Major (RSM) and loved ones were present to say good-bye. Tears were flowing, hugs were exchanged, as well as lots of handshakes and best wishes. Our families were nervous because of the terrorist attacks of 11 September 2001. We had only been able to tell them that we were off to South West Asia (SWA). We assured them that we would be safe and we implored them not to worry; all would be well and we would make contact at the first opportunity. The bus departed for Greenwood at 1300hrs.

CFB Greenwood was in high gear. They wanted to know who we were; where our doctors and our needle books were; who was in charge; had we settled our pay; and did we have additional passport photos. We had all the answers. This was not our first trip. Once finally settled in the barrack block, we went for supper and a couple of pops.

On 27 December, we tracked down the Wing Logistics staff in charge in order to ensure that our equipment and kit were on the first departing aircraft and to inquire about our departure time. We were told to meet at the mess in the afternoon as the Chief of Defence Staff (CDS) and other very important persons (VIPs) would be bidding us farewell and wishing us all a safe journey. We were also informed that we would be flying out on an airbus later in the evening with a stay over in Zagreb, leaving Croatia the morning of 29 December.

It was dark by the time we arrived in SWA at around 1900hrs. Nonetheless, it was still hot and the humidity had our clothes sticking to us. It was a big contrast from Greenwood which had just received a 150 mm snowfall. Moreover, our new location was nothing like what we had imagined. It appeared to be very modern and well designed. Everything looked shiny, especially the vehicles, possibly because it had not rained in five years, at least according to the locals. Once we cleared customs and got our luggage loaded into rental vehicles, we boarded buses to take us to our new home. For some, this would be home for the most part of rotation (Roto) 0.

Following a two-day acclimatization period and numerous meetings and o-groups, (early spring temperatures were approximately 14 degrees Celsius in the morning, rising to 25 degrees Celsius during the day) all SET members were straining at the bit to get started. Notably, engineers have deployed on every mission Canada has taken part in, designing and building all manner of camps and set ups, either using existing infrastructure (no matter the condition), setting up deployed assets, having new buildings built, both short term and long term, or a combination of the aforementioned. The timeframe authorized for this particular venture was 59 days in theatre. We had already lost two valuable days getting acclimatized. The SET Sergeant-Major was getting antsy.

Our team in theatre totalled fifteen military members: three Captains; one master warrant officer (MWO); four warrant officers (WOs); four sergeants (Sgts); one master corporal (MCpl); and two corporals (Cpls). The Site Activation Engineering Officer, a captain, was not from 1 CEU, and two WOs were augmentees from CFB Kingston and CFB Petawawa. The designing trades within this team were as follows:

> Heavy Equipment (042 Warrant Officer); Refrigeration (641 Sergeant); Electrical Distribution (642 Sergeant); Electrical Generating (643 Sergeant); Plumbing & Heating (646 Warrant Officer); Water Fuels Environment (647 Warrant Officer); Construction/Survey/Drafting Technicians (648 Sergeant, MCpl & Cpl); and a Communication/Signals Specialist (Warrant Officer). In addition, to ensure self-efficiency, each SET was assigned its own Resource Management Clerk (836 Cpl).

Our attitudes and approaches were dictated by our engineer hat badge: we were in a hurry and we had a lot of work to do in a very short period of time. We were the proverbial eager beavers. This deployment was not a holiday to golf, swim in the pool or shop at the souk. The hotel, daily per diem and rentals were costing the Crown significant amounts of money and we had to get this camp set up and running soonest. We travelled the forty-two kilometres from our hotel to our work destination early every morning via rental vehicle and we returned to our lodgings late in the evening. We worked seven days a week with a day off every three weeks or so. Time, however, continued to fly by. We remained far too involved in our daily assignments to think of anything else.

Once settled into two temporary office trailers, our daily work load included: surveying the existing site; finding the utilities; producing site plans; investigating and writing product and building specifications; all facets of design required for refrigeration, electrical distribution and power generation; and water and waste management for the operations site kitchen and dining facilities, the accommodations and numerous other facilities that enhance quality of life and morale of our deployed members. Additionally, there were daily reports to complete, personnel evaluation reports (PERs) to write and an endless number of meetings to attend.

This was 3 SET's first operational deployment. We had complete confidence in our ability to accomplish the task. Many of us had previously deployed on numerous occasions and we knew our jobs well; we also knew our limitations. Between the fifteen of us, we had experienced approximately eighty missions. For our newer members, this deployment provided a golden opportunity to demonstrate their trade skills and knowledge in designing a camp capable of housing approximately 150 personnel. Of course, however, there was some nervousness as with any project but, as a team, we worked things out and helped each other to overcome any and all obstacles.

Before we knew it, however, we ran into the proverbial brick wall. It was soon discovered that the existing utilities would not meet our needs for 150 personnel. We had to quickly change our plans to support a self-contained camp. The new plan would include many large diesel generators, waste management systems, above ground complex electrical distribution panels and a water system capable of handling all needs including fire fighting safety. We wondered where to get the new equipment and if the additional costs had been approved. Many other issues were raised every day and there were long delays while waiting for answers. We received directions from a multitude of command structures, J3 Ottawa, Florida, the contingent CO, the site engineer, and the logistics officer (Log O), to name a few. Indeed, the boundaries were complex and constantly changing for this team.

Designs and layouts were all but thrown out each week. Frustration and strained morale quickly became an issue for the Sergeant-Major to keep a close eye on. To add to the complexity of the situation, it seemed to the SET designers that after every o-group the number of troops to be housed increased. Because of the projected growth of the camp, additional infrastructure was being added to our list of things to do and consider. Moreover, with each additional group to be housed, generators had to be increased in size. This growth, in turn, required larger electrical cables to be run, coupled with larger water and sewer lines.

It quickly became clear that we were actually designing a small town. The buildings included: a gym; a mailroom; numerous supply and storage facilities; a petroleum, oil and lubricants (POL) point sewage lift station; an internet café; guard shacks; aircraft maintenance buildings; a very large operations centre; a kitchen and dining facility; an aircraft photo development building; and a headquarters complex and offices. This was certainly not going to be a tented camp or a Relocatable Temporary Camp (RTC) layout as we had originally been led to believe, moreover trained to design. Rather, the new plans called for hard wall constructed buildings, each specifically designed for the various occupants to perform their daily assignments, and a vast complex of accommodations.

The wind was all but knocked out of us when we were advised that CFB Trenton would be sending their Tactical Airlift, with approximately 150 (or more) personnel, along with Hercules aircraft. These additional requirements were to be included into our design and layout.

We worked even harder and longer hours than before. We overcame many obstacles. Eventually, we succeeded. Camp Mirage, as it is called, by the end of Roto 0, was capable of housing a 550-person sustainment camp for our troops in Afghanistan.

To state that our environment and site conditions were harsh is an understatement. No one had any experience working, or designing infrastructure to be capable of functioning, in the 40 to 55 degree Celsius range. We were in a desert with little or no room to manoeuvre due to the size of our final operations and accommodation layouts. These extreme high temperatures are non-existent in our Canadian codebooks and manuals. We were breaking new ground with our design and our electrical needs were enormous as almost each room required an air conditioner, which meant more and more generated power complete with larger manufactured electrical distribution cabinets and very large power cables.

Furthermore, the soil was as hard as granite. Although mostly sand, it had been sun baked and wind swept for thousands of years and had thereby solidified into a dense mass. We soon discovered that our work schedule and critical path for digging the utility trenches, which we had estimated at three days, was unrealistic. It took us three weeks. In order to save valuable time and money, we placed the three utilities in the same trench. We designed waste lines on the sloped bottom with water and electrical runs on either side. It must be mentioned that there were an additional dozen or so construction engineer tradesmen in theatre from an Airfield Engineer Flight deployed for Roto 0. This group was commanded by a captain and his WO. They were tasked with maintaining the infrastructure. After much discussion, a deal was struck and most of these tradesmen provided the bull-work in laying the water lines, sewer lines, setting up the kitchen and two dining halls, as well as a multitude of other tasks. It was evident that further engineering support from Canada was not going to be forthcoming.

The SET faced new obstacles each day. For example, not only did our design encompass North American equipment, we needed European power as well. The buildings being constructed required a European power supply; conversely, our kitchen and dining halls needed North American power. The two are not interchangeable. For instance, one day, after modifying his electrical cord on his radio, a section head finally thought he could listen to some rock and roll. His radio went up in a cloud of smoke. Another example is of alarm clocks running about two hours late during a 24-hour period since using European adapters does not change the electrical cycle from 60HZ to 50HZ. Certainly, prior to any overseas mission, all personnel should be advised, or reminded, of the difference in voltage and cycles between Canada and the host nation, thus ensuring expensive equipment is not needlessly damaged.

By February, one of the main morale issues for our SET team – which must be considered whenever units deploy as part of an activation team – was the length of the tasking message. As previously mentioned, we were on a "Temporary Attached Posting" for 59 days, which predictably had us leaving on, or about, 26 February 2002. It became evident early on, however, that we would be in theatre for quite some time. In fact, in the end, the last couple of members to leave had all but completed the full six months of Roto 0. Yet, although deployed members are entitled to Rest and Relaxation (R&R) breaks and a home leave travel allowance (HLTA) during a six-month deployment, 3 SET was not granted this luxury. This disparity became a huge morale buster and was the hardest obstacle for many of us to overcome. Almost daily, others were booking flights and making holiday plans for their HLTAs while our team could not enjoy this privilege. As a small team, we did not have the resources to allow members to depart for upwards of three weeks, nor did our home unit have additional teams available to replace us.

On, or about, day 62, we received a new message: "Extension to the Temporary Attached Posting." Now we had no end date. Watching our fellow engineers and other trades personnel leave for home and seeing them return rested and recharged had a grave impact on us. Many are still bitter about this perceived injustice. Notably, we received a ten-day special leave pass when we returned home. This comparatively small gesture, however, did little to compensate for the lost benefits during deployment.

By March, one of the many dilemmas that our team faced was associated with the high cost of constructing the camp. There were also governmental rules that we had to follow. As with any major construction project, we had to deal with Public Works Government Service Canada (PWGSC) to submit tenders for contractors to bid on. In this instance, the PWGSC representative had to fly in from Germany, as this was the nearest office. Luckily for us, the woman assigned to our group was exceptional. Not only was she well versed in all the procedures we were required to follow, she also offered some suggestions that we had not considered. One idea that she put forth was to accept two different contractors to build the vast number of structures required for the camp. This distribution of labour would drastically cut down on delivery time and would be a good means of competition between the groups.

This system ended up working well. Nonetheless, all target dates were missed by several weeks. The new system did not afford enough time for our designers to visit the manufacturing sites in order to complete quality assurance and control and rectify flaws before the delivery of the dozens of components required to put our buildings together. For instance, a single forty-personnel accommodation building was made up of six large sections. We had thirteen forty-personnel accommodation buildings being built, in addition to several other structures. Scheduling the deliveries for these materials alone became a major tasking for our Resource Management Support (RMS) clerk. Paperwork had to be filled out for the host nation at least twenty-four hours ahead of each delivery and separate paperwork was required for each contracted tradesman to enter the base. Passports, work visas, and the like had to be verified time and again. If there were any errors in typing or spelling, there was no entry permitted until the paperwork was redone. New paperwork meant that the whole process needed to be restarted from the beginning which resulted in a forty-eight hour delay. Errors included any typos or misspellings to the driver's name and nationality. Notably, it was difficult to ensure correctness when all the information was acquired through phone calls and there was a constant language barrier.

Despite these obstacles, things were starting to happen quickly and morale was lifting. We were now resigned to the fact that we would be in SWA for about five months. We were also told that we now had to incorporate a National Support Element (NSE) Headquarters for an additional forty-five personnel. There seemed to be no end to the size of our town.

National Defence Headquarters (NDHQ) approved the use of RTC assets from Italy for our kitchen and dining facilities. A few of us had some experience setting up Weatherhaven buildings, but collectively we had very little knowledge of erecting a kitchen and dining facility. Arrangements were made to have a civilian representative come to our location to provide the much needed technical assistance. This gentleman arrived on time with expectations that the RTC components would arrive the following day. Unfortunately, once again, there was a glitch somewhere and the sea cans were stuck on the dock and would not be released for another three weeks.

After much paperwork, we eventually started to receive these extremely heavy sea containers. Our new problem became offloading the canisters. We did not have a forklift capable of moving them. We quickly resolve this problem by using the two forklifts available in unison and we were able to end the day on a positive note. We subsequently sent out a request to the other units asking for any and all available personnel to assist us in erecting these large structures. It must be noted that although it was March, the temperature was so high that we could only get about twenty minutes of heavy work done before needing a ten minute rest break to replenish our body fluids. The average temperature in March was 25 degrees Celsius first thing in the morning, quickly climbing to 30-40 degrees Celsius by mid-day. Interestingly, the locals found it cool and could not understand how we could work wearing only T-shirts.

April arrived and we could begin to see the end. Some of the rules had changed, however. Since we were building this camp on an existing air base, we had to follow the established base guidelines and regulations. Although the paperwork had been properly filled out, with all the required information and submitted at least twenty-four hours in advance as required, and there were tractor trailers at the main gate ready to receive the shipment, the host nation advised us that they would no longer allow deliveries to be made on Thursdays or Fridays. The drivers, the contractors and our own RMS clerk had not been forewarned and were perplexed as to why they could not complete the delivery. They wondered if there was a problem with the paperwork or if some technicality had been overlooked. Such issues, which seemed to happen on a regular basis, were just more of what seemed like daily events designed to slow down our progress. In the end, we managed to have our kitchen and two dining halls set up with a contractor building the massive air-conditioner required just to keep the kitchen component cool enough so that the cooks would not pass out from the heat.

We then had to wait for word on the status of our electrical generators. The question was whether to rent, purchase or acquire assets from existing stock. These issues were being raised on a daily basis between our Site Activation Officer and higher headquarters. Finally, a decision was made. We would both rent and use generators from existing stock in Italy.

We eventually got a contractor to level off the sites for our infrastructure, utility trenches and communication systems. Two other contractors were building the concrete pads and footings for the numerous buildings. Since our lift station was installed, we had yet another contractor building the two holding tanks for water and sewage. Notably, we had not considered that the waste (human) would cook into a solidified lump during the extreme daily heat. Thus, once the camp was set up, many standing offer agreements had to be amended. For instance, the contractor who was hired to remove waste had to shovel out the holding tank instead of pump it out as originally planned. There was also an electrical contractor laying electrical cables and installing distribution panels. All these contractors had to be carefully followed to ensure that we received what we designed and what we had paid for.

Language and cultural barriers further frustrated the growth of the camp. The most common expression from the locals was "Inshallah," which means "God willing." Arguably, however, a more accurate translation would be "perhaps" or "maybe tomorrow."

By the end of April the camp was well underway. The Engineers were nearly exhausted, but we knew that in six to eight weeks we would be returning home for a much deserved and needed rest. The personnel from 3 SET had gone about their respective jobs with a cool efficiency and an unerring sense of dedication that ensured the success of the mission. 1 CEU could not have found a better team to tackle this nearly impossible task if they had canvassed the entire country. The many Canadian Forces (CF) members and dignitaries who have stayed in, or have deployed to, Camp Mirage during Roto 0 and since then, are a testament to the hard work and dedication that 3 SET put into the construction of the camp. They all owe a debt of gratitude to a very small group of very dedicated engineers who got the job done in order to establish one of the finest camps ever designed and built in a very harsh land.

Following this deployment, an in-house investigation and a review of our standard operating procedures (SOPs) was conducted in order to ensure that other groups would not experience the same obstacles that we had encountered. One important lesson learned during this mission pertains to the perception of genders in foreign cultures. We were in a predominantly Muslim country where men do not take orders from women. This gender discrimination is their reality and must be considered prior to any deployment. Those locals with higher education and travel experience understood that Canadians have both male and female officers and fully supported most directions from men and women. They, however, represented a minority. The vast majority of locals would not accept that a female was in charge. Careful consideration of these cultural nuances must be taken when a Theatre Activation Recce is stood up, particularly when time is a factor and local support and labour is required. It would also have been very useful for a senior member from 1 ESU to have deployed on the recce. This individual could then either deploy with the implementation SET or could be used as a main point of contact to clarify design issues. The Canadian Embassy also proved to be a good place to find information about the local culture as well as various lists of contracting companies, some of which even had headquarters in Canada.

Although various standard camp designs exist for the CF, depending on the length of the deployment, the Air Force must adapt to, and accept, the fact that they may be in modular tents, RTC Weatherhaven structures or ISO accommodations. It is extremely difficult to design and build a camp where it is perceived that standardized CF regulations do not apply to aircrews. Claiming that they cannot sleep at night due to aircraft noise during take offs and landings is unacceptable. Every other trade adapts to the environment. Yet, repeatedly, special arrangements are made for aircrews and this type of favouritism severely affects those tradesmen and tradeswomen who are not considered special. The army, for instance, has always accepted whatever is made available to them and small comforts have been realized during their tours.

Since our return, many questions have arisen. People have wondered about the length of time it took to build the camp and the overall cost to the Crown. In the end, all the delays and conflicting information which led us all over the place with our designs have been added to our "Lessons Learned" notebook. Yet, as with any deployment, the difficult moments, the stressful timeframes and all the daily whining have since faded and only our successes have remained fresh in my memory. Although the deployment was difficult, our accomplishments during this 59 day (6 month) tour were tremendous and we are all proud of our accomplishments.

CHAPTER 2

WALKING THE LINE AND KEEPING THEM HAPPY: THE LEADERSHIP EXPERIENCES OF CPO1 ERIC CARBERRY ON OPERATION APOLLO, ROTATION ZERO

Craig Leslie Mantle[1]

When the terrorist attacks on the World Trade Center occurred on 11 September 2001 (9/11), Her Majesty's Canadian Ship (HMCS) *Iroquois* was at sea conducting vigorous training exercises. Once the extent of the immediate damage and death toll were more fully appreciated by the Government of Canada and the Canadian Forces (CF), the ship quickly returned to the dockyards at Halifax with orders to stand in readiness to deploy to New York City in order to provide humanitarian assistance and to aid in disaster relief efforts there. In the end, however, this tasking never materialized, although the ship's company was soon given another that would take them much further abroad than continental waters. A month later, in mid-October, *Iroquois*' sailors were informed that they would soon be deployed to the region encompassing the Persian Gulf and the Gulf of Oman in support of the United States' recently-declared War on Terror. Rather than having the usual lead-time before a scheduled tasking to ready the ship and to attend to personal matters, whether at home or elsewhere, all hands were informed of their imminent departure on Operation (Op) Apollo only one week before sailing, thus leaving little time for either.[2]

After learning of this upcoming deployment, some sailors who had earlier submitted their releases well before 9/11, but who were now simply waiting for their paperwork to be processed, indicated their willingness to stay with the ship and to sail overseas. For these individuals, the mission came first above all else and any personal preferences would have to wait until their return to Canada. Seeing this level of professionalism and commitment immediately heightened morale and reaffirmed for many that a team spirit still prevailed. Because time was very much at a premium, *Iroquois*' coxswain, Chief Petty Officer First Class (CPO1) Eric Carberry, directed the various department chiefs to make certain that all of their sailors spent at least some time at home with their family and friends.[3] Preparations for the mission were given a clear priority, but a conscientious effort was also made to ensure that personal matters were attended to prior to deploying.

The day that the *Iroquois* and two other Canadian vessels sailed from Halifax, family and friends, in addition to well-wishers from the city and other local areas, lined the dockyard and waterfront along the departure route. Of particular note, a strong contingent of Legion members had fallen in with full colours to bid the sailors fair winds and following seas. CPO1 Carberry and others felt this display to be a great mark of respect, especially coming from these dedicated individuals who, decades earlier, had themselves left Halifax to venture into the unknown. The people of the city have always shown their support for those heading into harm's way and this occasion was indeed no different. All aboard ship felt extremely inspired and motivated by the sight for they knew, at the very least, that they were in the thoughts of those left behind.

Even before starting for the Middle East, however, the crew of the *Iroquois* had coalesced into a relatively cohesive team and functioned well together for they had just returned from demanding exercises at sea. In order to ensure full operational effectiveness whilst deployed, the Navy decided that the ship would sail with a full complement of sailors. Shortfalls in personnel were quickly made good, although a few of those posted-in arrived only a day or two before sailing, thus leaving little time for team building. Despite these changes though, more than 60 per cent of the ship's company had already served together, so these disruptions did not negatively impact the *Iroquois*' overall functioning and efficiency to any appreciable extent. The lengthy transit between Halifax and the two Gulfs provided the time to fully integrate the "newcomers" into the pre-existing team. Work-ups not only ensured that the ship was operationally ready, but further melded the crew into an effective whole by reinforcing trust and fostering cohesion amongst all onboard. Although Carberry was not compelled to actively ensure that a strong team environment was created, as would have been necessary if the majority of crewmembers were unknown to one another and had not served together in the recent past, he still monitored their progress on this account and encouraged unity whenever the opportunity arose.

Despite the benefits accrued to cohesion during transit, this period was extremely taxing on all as the sailors spent the majority of their time training for the unknown. Upon leaving Canada, no one onboard really knew when they would return or what specific challenges they would face once in theatre. Much effort was consequently expended in training for all types of possible scenarios, such as those that might include nuclear, biological or chemical threats. Along with the remainder of the command team – the Commanding Officer (CO) and the Executive Officer – CPO1 Carberry ensured that the sailors were kept busy and that their work had a clear and definable purpose. He reasoned that being occupied with an intense training schedule would keep their minds active, thus lessening some of the fear and stress that they might otherwise have felt if they were left idle with little to do but ponder their possible futures. Although demanding much of the crew, he also took pains to ensure that they were not pushed too hard, too far or too fast, for he believed that overwork would only lead to frustration and resentment that would, in turn, ultimately affect performance. A fine line had to be drawn between too little and too much and his ample experience allowed him to judge this path accordingly. In the end, the crew arrived in theatre after an exhausting and busy transit, confident in their many skills and working well together as a fully integrated team.

From the time of their arrival in their area of responsibility in November 2001 until January 2002, the *Iroquois* shadowed and provided protection to a US Amphibious Ready Group (ARG), composed of the USS *Bataan* and other vessels, that was responsible for landing US Marines and then supporting their ground operations inland.[4] These initial months were extremely busy and left the sailors with little spare time and thus the opportunity to be bored; the excitement of being deployed had not yet worn off either.

Once departing company with the ARG early in 2002, the *Iroquois* began conducting both leadership and maritime interdiction operations in which members of the crew boarded foreign vessels in search of Al-Qaeda and Taliban terrorists and their weapons. Patrolling the same waters over the course of a number of months eventually became both monotonous and tedious. To counter the insidious effects bred by such familiarity, the command team permitted and encouraged a number of events that provided necessary distractions.[5] Barbeques were held frequently on deck and Sunday became a "No Day," in which very little was done aboard ship, excepting those duties that were absolutely necessary or dictated by operational commitments. The novelty of using CO_2 cannons to launch apples at enlarged photographs of the perceived enemy was not lost either!

With some forethought and effort, CPO1 Carberry, the department chiefs and other crewmembers did the best that they could with the resources onboard to ease the tedium. For him, the challenge was to "keep morale going in a sometimes mundane environment" and to keep the monotony at bay.[6]

Encouraging morale was not solely a reflex action to a perceived problem. From start to finish, the entire deployment was marked by attempts to keep everyone committed, focused and in good spirits. At Christmas time, for instance, some of the crew organized themselves into a choir that started practicing hymns and carols well in advance. Around Christmas Day, the *Iroquois* sailed through the ARG and, once alongside other ships, the choir began to sing over the loudspeakers. In some instances, these "hailings" were returned in kind. Such impromptu concerts did wonders for morale as the message traffic between ships confirmed the following day. Other activities included a decorating competition between the various departments and a "Secret Santa" game in which messmates drew the name of another crewmember and then offered him or her a gift. Importantly, they also decorated a Christmas tree – really just an improvised assemblage of metal pieces – with glow-sticks. Attempting to make the holidays at sea as similar to those enjoyed at home surrounded by family and friends was extremely important to Carberry and the crew. Everyone's collective efforts in this regard helped displace some of the sadness naturally felt by being separated from loved ones at such an important time of the year. Even a few of the nay-sayers, who initially would have nothing to do with such activities, later participated for they soon realized how "into" the spirit of the season all were becoming and how much fun and companionship they were missing by isolating themselves. To continue a long-standing tradition aboard the *Iroquois*, lobster was provided during the Christmas period, much to everyone's enjoyment.

Morale was also maintained throughout the deployment by the support that *Iroquois*' sailors received from all parts of Canada. Thousands of letters, postcards and banners were proudly displayed on the ship's bulkheads and were a constant reminder that those back home had not forgotten them. On many occasions, sailors who were feeling down and unhappy stopped to read a passage from a complete stranger, only to remember why they were overseas putting themselves in harm's way. Such displays of support were extremely motivating and left the crew with a profound sense of accomplishment. In Carberry's opinion, these greetings from Canada probably "had more to do with our high morale than did the command team and department heads and chiefs."[7]

When the *Iroquois*' responsibilities changed in early-2002, a host of new leadership challenges presented themselves that required novel and inventive solutions. Prior to 9/11, according to Carberry, sailors throughout the Navy understood their duties and how to accomplish them both safely and efficiently. In sum, sailors sailed, end of story. When deployed on the first rotation of Op Apollo, however, the entire ship's company was confronted with a number of new duties with which they had very little, if any, experience or preparation. For instance, after Christmas, the *Iroquois* sailed to Dubai for a port visit where the ship's engines were replaced over the course of a week or so.[8] When alongside, security and force protection became an extremely high priority, whereas such was not always the case. In order to ensure that the ship and port remained safe – they of course presented tempting targets – many of the sailors were forced to search incoming vehicles that entered certain facilities at the dockyard, a task that many had never performed before or had any familiarity with. In addition, many had to stand watch with loaded weapons whilst in harbour and to counter any threats according to the established Rules of Engagement (RoE).

As might be expected, many had legitimate and immediate concerns about their new responsibilities. Few, for instance, knew where and how to look for vehicle-borne explosives. Their earlier years in the Navy had certainly not prepared them to deal with such threats. In response, CPO1 Carberry and the command team partnered *Iroquois*' sailors with members of the Military Police (Canadian) for their first few rotations on duty in order that they might have their questions answered, thereby instilling confidence in their abilities. After receiving such assistance, the sailors eventually felt sure enough to assume these difficult and demanding tasks on their own. In this instance, outside expertise was relied upon to close a clear gap in knowledge. Similarly, after receiving direction from the CO, Carberry and the department chiefs made sure that everyone aboard ship, including the officers, fired their primary weapon at least once a week – some used the standard CF C7, while others employed .50 calibre machineguns – in order to make them more comfortable with engaging enemy targets and defending themselves. These practice shoots, which were conducted primarily at sea, were duly recorded so that the captain knew at any one time how many crewmembers had received this training and how many still remained deficient. To supplement these exercises, the CO and coxswain briefed the RoEs every day to the entire duty watch, again ensuring that all possessed a certain level of comfort and familiarity with these rules if they were ever forced into a compromising situation. All in all, the command team and senior leadership made a conscientious effort to further train their crew during the deployment in order that they might gain additional skills suited to specific operational requirements and, more importantly, confidence in their own abilities.

CPO1 Carberry relied heavily on certain members of his own crew who possessed certain expertise to further mitigate the stress caused by the imposition of these new taskings. A number of the ship's company had incidentally served in the Army before switching uniforms; many of them were ex-firefighters. For them, weapons handling and RoEs were "old hat" and something with which they were intimately familiar. Those sailors who lacked ability or comfort with their assigned weapons were oftentimes paired with such individuals who duly provided the necessary training, guidance and advice. As might be anticipated, the ex-Army types found the chance to get behind a weapon again extremely satisfying, so much so that they often volunteered for extra "shit jobs" in exchange for the opportunity to "help with the 50s."[9] Their willingness to assume extra duties was to Carberry a sign that the morale and team spirit so essential for success were sufficiently well advanced aboard the *Iroquois*. By utilizing those individuals who really wanted to help train the remainder of the crew, other sailors did not have to be pulled away from their primary duties in order to provide instruction, thus easing their burdens in turn.

In due time, after spending six months at sea, either protecting other vessels from attack or searching for the weapons of terrorism, human or otherwise, the *Iroquois* returned to Halifax. In the estimation of the ship's coxswain, the deployment had been an overwhelming success. Taskings had been completed with an exceptional level of professionalism and the experience gained by everyone would serve them well in the new environment in which the Navy, and the rest of the CF, would now operate.

While initially quite high given the excitement of being deployed, the crew's morale sometimes waned owing to the monotonous and tedious nature of their work. To counter these effects, CPO1 Carberry, the department chiefs and the remainder of the command team strove to offer as many diversions as time, space and resources would allow. Throughout the deployment, the stress induced by new assignments forced the senior leadership to develop methods of limiting its pervasive influence. To this end, they capitalized upon resident expertise in order to train those who required extra instruction and strove to offer a number of memorable distractions. During his time aboard HMCS *Iroquois*, CPO1 Carberry ensured the welfare of his subordinates through whatever means possible and by walking the fine line between pushing them too hard and not enough.

ENDNOTES

1 This chapter is based on an interview between CPO1 Eric Carberry and Craig Leslie Mantle, which was conducted in Halifax, Nova Scotia, on 21 July 2006.

2 Roto 0 included HMCS *Iroquois*, *Halifax*, *Charlottetown*, *Toronto*, *Preserver* and *Vancouver*, and, at the extremes, lasted from October 2001 to May/June 2002. Commodore Drew Robertson served as the Task Group Commander. See Richard H. Gimblett, *Operation Apollo: The Golden Age of the Canadian Navy in the War Against Terrorism* (Ottawa: Magic Light, 2004), 80.

3 The department chiefs under Carberry's command, all of whom held the rank of Chief Petty Officer Second Class (CPO2), included the Supply Chief, Deck Chief, Marine Systems Chief, Combat Systems Chief and Air Chief.

4 For additional information on this phase of their deployment, see Gimblett, *Operation Apollo*, 47-8.

5 Even before the *Iroquois* set sail from Halifax, the command team took concrete steps to mitigate some of the stress inherent in overseas deployments. To this end, they either "bummed or bought" (interview quotation) exercise equipment for an onboard gym. Given that many sailors were extremely health conscious, such a facility was an absolute requirement and the lack of one would have negatively affected morale. By the end of the deployment, many of those who had not habitually exercised in a gym setting before had become quite interested and were using the available equipment with some frequency. Having a place to exercise and socialize informally also staved off the sapping effects of both boredom and monotony and contributed to a greater sense of community.

6 Interview quotation. To further counter the effects of boredom, numerous talent contests and bingos were hosted aboard the *Iroquois*. The crew also followed the Olympics

with interest and took great pleasure in watching both the Canadian men's and women's hockey teams win gold in Salt Lake City.

7 CPO1 Carberry to C.L. Mantle, 7 Sep 2006, via e-mail.

8 Incidentally, when the engines were being changed, many sailors from outside the engineering department volunteered to assist with those tasks that did not require technical expertise, such as helping with certain procedures, holding a light, fetching tools, etc. This allowed the engineers some time away from their jobs and the chance, which everyone else had, to enjoy the port city. Such willingness to help, Carberry believed, was a very good sign that earlier team building efforts had been successful and that both morale and cohesion were on a steady footing. The sailors who willingly offered their help did so at the expense of their own personal time and enjoyment. As earlier, when in Halifax before sailing, Carberry told the department chiefs that once all of their required work was complete, their people could (and should) go ashore. A conscientious effort was made by the command team to give the sailors as much leave as possible when the opportunity so presented itself.

9 Interview quotations.

CHAPTER 3

OPERATION APOLLO THROUGH THE EYES OF THE BATTLE GROUP RSM, FEBRUARY-JULY 2002

J.A. Comeau

The terrorist attacks on the United States (US) on 11 September 2001 (9/11), have forced us to re-evaluate war. Traditionally, only Special Operations Forces, with their surgically precise tactics, have been the predominant force to engage terrorists. However, 9/11 transformed this reality. This terrorist attack in many ways changed the concept of how the West wages war. The new emphasis on asymmetric warfare created an environment that demands that conventional forces be engaged against terrorists. Conventional forces, however, would need to adapt in order to survive on the new battlefield. The Canadian Forces (CF) was not immune to this need for transformation and it needed to acclimatize quickly in order to survive and contribute positively to the new defence environment. The following is an account from my perspective as the Regimental Sergeant-Major (RSM) of the 3rd Batallion, Princess Patricia's Canadian Light Infantry (PPCLI) Battle Group on Operation Apollo, from February to July 2002. I must remind all who read this chapter that there are always two sides to the story and I forewarn you that I am about to give you just one side of it.

MISSION

Most Canadians were introduced to Osama Bin Laden on 11 September 2001. By that time, they had already forgotten the attacks on the World Trade Center in 1989 when a truck bomb was planted in the parking lot several stories below the structure. North Americans often allow themselves the luxury of thinking that they are immune to attack because they live under a formidable security umbrella. Indeed, a mere few hours before the attack on the Twin Towers, the general population probably did not even think that such an act of terrorism would have been able to be carried out in North America. In reality, however, as 9/11 has shown us, our continent is not impenetrable to attack.

Prior to 9/11, the 3 PPCLI Battle Group was being formed to go to Norway on a North Atlantic Treaty Organization (NATO) exercise, scheduled to commence in February 2002. As early as 1 April 2001, we had been assigned as the Immediate Reaction Force Land (IRF (L)). I, as the RSM, was in the process of ensuring that there was a name to every established position so that the Commanding Officer (CO) would have a full complement of soldiers to conduct the exercise in Norway. At the time, the Light Infantry Battalions were on the chopping block so the Battalion was undermanned by at least a rifle company's worth of soldiers. The company command structure still existed within the Battalion, but we did not have enough soldiers to man all three companies. Consequently, the CO decided that "A" and "B" Companies would be fully manned and he would go to the Brigade Commander to get another company from outside the Battalion. 2 PPCLI was tasked to provide the third company. The 3 PPCLI Battle Group was thus complete to deploy to Norway for the NATO exercise. The Battle Group included 3 Rifle Companies, Combat Support Company, Administration Company, Combat Service Support Company, Engineer Squadron and attachments to fill the headquarters.

As I was driving to work on the morning of 11 September 2001, I heard on the news that an airliner had struck one of the Twin Towers of the World Trade Center. By the time I was going through the main gate at the base the second airliner had struck the other tower. An eerie feeling began to fill the pit of my stomach. During morning parade, I warned the Battalion that the world as we knew it had just come to an end. I told them that I would be surprised if we were not soon somewhere in the world in retaliation of the strike.

The events of 11 September and the months that followed are well documented. For the 3 PPCLI Battle Group our exercise focus in Dundurn, Saskatchewan changed from preparation for a NATO exercise to hunting Taliban and Al-Qaeda.

With this new objective in mind, the original plan had the full headquarters organization coming together so that we would be able to gel as a team and not have to work together for the first time in a theatre of war. Much to the discontent of the CO, however, this objective was not met. We managed the headquarters by having designated staff officers double-hatted as support staff.

Owning the night was one among many priorities that the CO had set if we were going to fight a war against terrorism. For this and other tasks, we would need the right equipment. The 3 PPCLI Battle Group was fortunate to have among their ranks former members of Joint Task Force 2 (JTF 2). These individuals were a tremendous help in identifying the best equipment for the upcoming fight. The following is a list of equipment that was to be brought into the unit prior to the deployment to Afghanistan: PAC 4 (laser designator); PAC 5 (laser designator); Night Vision Goggles (NVG) (one per man); level 3 body armour (one per man); Glint Tape (the system did not produce the tape, a former member of JTF 2, now within the Battle Group, had to use his contacts with Special Forces on the ground in Kandahar to get enough tape for the Battle Group); Infrared Strobes; and a C8 Carbine rather than a C7 Carbine with an issue of 10 magazines per man. These acquisitions and other initiatives ensured that the soldiers got the message: we were not going on a peacekeeping mission; the focus was definitely war-fighting and we were committed to mission success.

One of the first dilemmas that the Battle Group had to deal with was the proper helmet mount for the NVG. Among international forces, it is said that the Canadian soldier can bring a solution to the table no matter what the problem. This case was no exception. A soldier of the Battle Group solved the NVG mount problem by removing the main mount off the head harness and screwing the NVG to his helmet. With no other suggestions on how to solve this problem, there were two acceptable ways of mounting the NVG onto the helmet. One was screwing the mount onto the helmet (which, according to experts, lowers the ballistic protection) and the other was to modify the head harness to fit the helmet.

The introduction to Multiple Integrated Laser Engagement System (MILES) that we borrowed from the Americans was also a valuable tool for our training. We soon realized that blue on blue was a definite possibility when manoeuvring large formations of soldiers.

To battle harden the soldiers, it was decided that we would not use arctic tents when staying overnight in the training area. On deployment to Dundurn snow was not on the ground. By the third day of training, however, snow had come to stay in the training area. Not using arctic tents proved to be a very valuable lesson to the Battle Group as it was soon realized how versatile and stoic the Canadian soldier could be. The soldiers understood why we were doing the training and being challenged by the elements added to the experience. For safety reasons, of course, the battle group always had tent groups close by and these were used on occasions when the temperature dropped significantly below freezing.

On our return from Dundurn, we heard that our likely mission would be with the International Security Assistance Force (ISAF) and we would be attached to the British Army. This meant that we were going to be a light support vehicle wheeled (LSVW) transported Light Infantry Battalion. This particular ISAF mission failed to materialize, however. Rather, in December we learned that the US was asking for the Coyote and its surveillance package to help in the defence of the Kandahar Airfield. A recce squadron from the Lord Strathcona's Horse (Royal Canadians) (LDSH (RC)) was designated as the unit that would provide that capability.

Consequently, the Battle Group now had a different focus and the Recce Squadron was added to the existing IRF (L) organization. To support the War on Terror, we were now going to be part of the US effort in Afghanistan. We would be attached to the only Air Assault Division in the world, specifically the 3rd Brigade Combat Team, "RAKKISAN," with a Headquarters Company and 3 Battalions of the 187th Infantry Regiment. It was obvious that our mission was not going to be a traditional peacekeeping mission. The soldiers knew that Afghanistan was going to be as close to war as the CF had been since the Korean Conflict. Knowing that the 101st is an airmobile unit, the vehicle infrastructure could be reduced as the Battle Group mission was only going to Kandahar as airfield security.

In preparation for our deployment to Afghanistan, we had a member of the Canadian Engineer Regiment who had been tasked to help the Mujahadeen in de-mining Afghanistan in the late 1980s come to talk to the Battle Group. His presentation on the culture and the mines that the Battle Group might encounter in Afghanistan was very well received by the soldiers.

When the Prime Minister came to visit us in Edmonton to announce that we were to be deployed to Afghanistan to fight in the War on Terror, we were again reminded that this was going to be a different type of mission than what we were used to. Of course, the visit had many layers of meaning, but the most important one was that the Prime Minister was actually taking the time to see us off to war. Another message was directed to the Canadian public: the government of Canada was actually doing something about the crisis that had occurred in North America, specifically, the CF was being deployed in the War on Terror. The government requested public support for the troops.

The Battle Group was sent home for the Christmas holidays. We were not told when we would deploy, however.

Once we were told that we were moving by C5 Galaxy transport aircraft all the packed pallets were prioritized and stored in a secure location. The US Air Force was given a window on getting the Canadian Battle Group into Kandahar. In hindsight, if we would have been told that the 101st Airborne Division was being deployed before us in January and that we were deploying after they were on the ground, then at least the Battle Group could have made an estimate for the departure date. This lack of dissemination of information is, in my opinion, where the JOINT aspect of our expertise was deficient. Moving a Canadian Battle Group using forecasted airlift is complex. In this instance, the task was further complicated because those planning the operation were not privy to the secure information which drove the airlift.

This period of uncertainty was a trying time for the Battle Group as some soldiers said their farewells to their loved ones on more than one occasion. For instance, I returned home three times before I finally left Edmonton. This was emotionally draining for my entire family. They would much rather have only said goodbye once. After the second time that our flight was delayed, soldiers were talking about not going home in order to avoid going through that emotional experience one more time. The leadership, however, ensured that most soldiers went home so that there would be no further complications should spouses have found out about the delay.

Since the Battle Group had not been together a single time before our deployment – and that point concerned me to no end – I felt that team building was going to be my biggest challenge. The other major concern that I had was the alcohol policy, especially given that we were going to be armed twenty-four hours a day, seven days a week. The latter proved to be a non-issue, however, as the US Army has a strict no-alcohol policy while on operations.

We were issued new CADPAT camouflage clothing immediately prior to our deployment. The uniform was what I used to ensure that everyone in the Battle Group looked alike. The 3 PPCLI Battle Group would be known as the "CADPAT Battle Group." I ensured that the word was sent out that no berets would be worn on the mission. The soft hat of choice would be the CADPAT floppy hat. Notably, this was not a popular choice among the rank and file of those with a different colored beret and the Command Teams of other Units attached to the Battle Group.

Once we were all on the ground in Kandahar we all looked like we belonged to the same organization and we were easily identified by our professional demeanor. My goal had been achieved. First impressions are always long lasting.

On arrival at the Kandahar Airfield, the advance party was given a piece of ground that would prove to be just big enough for our 844 man Battle Group. The US Camp Sergeant-Major informed us that he was not sure if the piece of ground had been cleared of mines. By the time I arrived, five days after the advance party had been on the ground, the bivouac had already taken form. The sanitary conditions were not up to par so the word was passed to the Canadian Camp Sergeant Major that we would adopt the "WainCon" (Wainwright Training Area Concentration) austere conditions standard. Within days, gray water pits covered urinals and sectioned outhouses were constructed with half-45 gallon drums. These conditions were a welcome sight to the soldiers, especially compared to the open pits that the US soldiers were still using. Hygiene and sanitation were my primary concern. Having experienced bad conditions before, I knew how important these issues were.

About a month and a half into the tour, everyone's morale was boosted when we moved to the prepared camp that we had constructed with a shower point, latrines and wash points. Living conditions had drastically improved. The addition of the chemical toilets, which was a familiar sight to the soldiers, was one of the most morale boosting items that arrived in Kandahar. The Battle Group even had a wash station at each chemical toilet sight. The morale of the soldiers had not been so high since their arrival.

Security was always a concern for the Battle Group. The Americans were allowing local Afghans to come on the base and do menial jobs. Both

Canadians and Americans escorted them. Through this action, our soldiers noticed that some of the locals were measuring distances and drawing diagram of the base. Those who were caught were not allowed on the base again. The CO took a firm stand on the issue of our Battle Group allowing Afghans on the base. We would clean our own areas. Some of the soldiers appreciated this decision while others with less operational experience did not realize why this decision was so important to the security of the Battle Group. It soon became evident to most, however, that the war in Afghanistan could quickly infiltrate into any area of life.

The Battle Group was declared operationally ready by mid-February and the soldiers occupied the defensive positions around the airfield. It was soon apparent that the situation had not much improved since the Marines had originally taken over the sector in the aftermath of 9/11. Some of the principles of establishing a strong defence had not been followed. The rifle companies therefore started the occupation by improving their positions. Moving out on the line and staying there was the priority of the Battle Group. Knowing that the soldiers were going to be on the line twenty-four hours a day, seven days a week fostered a sense of ownership within them and this became evident in the effort put forth to build their fortifications. The soldiers knew very well that this was not going to be a four-day defensive exercise where at the conclusion they would have to fill the trenches. The soldiers put their hearts and souls into the defensive positions. By the time the relief-in-line came to allow us to go on our first offensive operation, the US company commanders were suitably impressed with the work that our soldiers had done.

Having been told prior to our departure that we would be in a defensive role at the Kandahar Airfield, I knew that it was going to be a challenge to keep the soldiers focused on their mission. Once the CO relayed that the Americans were going to relieve us so that we could go and capture a mountaintop in support of Operation Anaconda, I was relieved for many reasons, particularly because I now knew that the Americans trusted us enough to allow us to fight side by side with them. In my opinion, the Americans gained this trust by watching us do battle procedures in order to rescue American soldiers in the Shi A Kot Valley. The level of experience and detail that we brought to the table was quite impressive. Our company commanders had the same experience/time in as the American battalion commanders. Our platoon warrants had the same experience/time in as the American battalion sergeant-majors.

Our soldiers too could see the difference in the leadership between Canadian and American officers and non-commissioned officers (NCOs). At the grass roots level, the Canadians had great confidence in their leadership. The soldiers of the Battle Group noticed, however, that American senior leaders were very inspirational orators. In contrast, compared to their American counterparts, when Canadian senior leaders came to address the soldiers they lacked charisma and energy. Canadians have been away from war for too long to be able to motivate soldiers for battle as the Americans do. The perception that the soldiers from our Battle Group held was that American senior leaders did not worry about what the press was going to say about their speeches. The Canadian senior leadership, however, showed some reluctance in front of the media.

Assets to the Battle Group included mapping and charting, Electronic Warfare, Intelligence, Reconnaissance Patrols and the CO's translator from another governmental agency. These assets proved to be very useful in all phases of the war.

OPERATIONS

Canadians were going on the offensive in a war zone for the first time since the Korean War. The Battle Group was to move from Kandahar Airfield to Bagram by C-17 Globemaster aircraft. Two Chinook helicopters were sent to Kandahar to pick up the advance party, which included Call Sign 0 and 9er Tactical. The Russian Occupation Forces had built Bagram in the 1980s. The Americans were now using it as a staging base for operations in southeastern Afghanistan.

As we were preparing for this offensive operation, I encountered an ethical dilemma. A soldier had approached his section commander and had informed him that because of his religious beliefs he could not participate in the upcoming operation. This seemed quite odd to me since the soldier was a volunteer and had essentially known what he was getting into prior to leaving Canada. I asked the company sergeant-major (CSM) what religion the soldier practiced. He answered Roman Catholic (RC). I too am a RC, albeit a non-practicing RC. I told the CSM to inform the soldier that the RSM is also an RC and that I was going on the operation to do my job for the kids of my kids so that they will not have to do what we are doing when they grow up. The soldier did not bring up the subject again. I think he was looking for a good reason why he should not go into battle, yet he could not find one.

Another issue that the CO and I discussed at length was how to fit the media into our first offensive mission. The media coverage of the occupation of the Kandahar Airfield was routine. The CO decided that we would bring the Canadian reporter with us as part of the Battle Group. This proved to be ground breaking in terms of fostering good media relations. With the reporter embedded into the Battle Group, the Canadian public and, especially our loved ones, could be intimately involved in the operation. As soon as the first couple of press releases were sent out from the field, everyone wanted a piece of the action. On one of the re-supply helicopters a Canadian Broadcasting Corporation (CBC) team arrived to cover the battle. Notably, there is one thing that I noticed about the CBC media team, they did not come well prepared.

When we arrived in Bagram to stage into the mountains, Operation Anaconda was winding down. The Americans were withdrawing from the valley that they had been fighting in for the nine previous days. We were to go and clear a mountaintop that had not been cleared. The Battle Group staff had named the hill feature "Whale Back" as the piece of ground protruded out of the earth as if it was a beached whale. During Operation Anaconda, mortar fire had come from this hill feature. Since it was on the other side of the valley floor, however, the Americans had not cleared the area. The Afghan National Army (ANA) had occupied the town in the valley and a small force had secured the helicopter-landing zone on the east side of the hill feature for the Battle Group. Yet, because the hill feature was not secure, the Battle Group was apprehensive about the landing going into Operation Harpoon. This was the first helicopter assault by a Canadian Battle Group in a war zone.

The most striking scene that I witnessed as I approached the airfield staging area was the padres giving their blessing to the soldiers prior to boarding. The one thing that the CO mentioned to me that he saw was that the soldiers were putting camouflage paint on their faces. The last thing on my mind was putting camouflage paint on my face as all the contingency plans were going through my head. It was an impressive site as the American Chinooks and US Marines Cobra Gunships moved out from the Bagram airbase to act as escorts.

As the helicopters landed on the "whale back," feature the anticipation of battle wore off as drills were executed and an all around defence was established. The remainder of the operation became routine, clearing bunkers and ensuring that the Taliban/Al-Qaeda could not come back to use the materials that we found.

On the first day of the air assault an incident occurred on the landing zone (LZ) that reaffirmed my faith in our soldiers. A young soldier was coming off the helicopter ramp and suddenly lost her footing. She fell and jammed her leg behind her back. The loadmaster on the aircraft went to help her off the ramp but she motioned that she would be able to do it herself. When I went to see her while she was being treated where she had crawled off the helicopter, I asked her why she had not stayed on the aircraft so that she would be able to get better care back in Bagram. She stated simply that she was carrying ammunition for the C6 machinegun and that she wanted to ensure that the gunner was going to get it before she would go anywhere. This confirmed in my mind that she had understood when I had emphasized to the Battle Group the importance of mission, buddy, self then everything else. I will always hold this soldier's action near and dear to my heart for she was an excellent example of the dedication that our Canadian soldiers demonstrated that first day of battle.

On our return to Kandahar Airfield, we staged out of Bagram. During the redeployment, the CO wanted to plan a rest and relaxation (R&R) programme for our soldiers. The original plan was to have the R&R at the Kandahar Airfield. After reassessing the living conditions and the level of operational readiness at the Airfield, the CO decided that it would be better if we had R&R outside of the operational area.

The US Brigade Combat Team (BCT) whose command we were under had an operational cycle that rotated Battalions in and out of the operational area. The R&R location for the BCT was in Pakistan. We, however, did not have permission to go into Pakistan so our R&R stayed autonomous to that of the Battle Group.

Dubai, which is located in the United Arab Emirates (UAE), was chosen because of the Canadian sustainment flights which were coming in from South West Asia (SWA). The flights came in on a regular four-day schedule. Moreover, this R&R location would allow the soldiers to be away from the operational area for a short period of time. As a precautionary measure, the CO required that the soldiers go in formed units. This way, if they had to meet us in an operational area after their return from R&R, then they would already be in a formed unit.

The R&R plan was being worked on as we were returning from Bagram. While we were still in Bagram, I was tasked with finding a suitable senior non-commissioned member (NCM) to man the leave centre in Dubai. With the blessing of the CO, it was decided that the senior NCM chosen would be someone who had previous experience at running a welfare facility. This experience would help to minimize the learning curve required for the position. This was particularly important because the soldiers were going to spend their three days leave in a country that was 96 per cent Muslim. Muslim laws in the UAE were intolerant of typical North American behaviours and, if our soldiers were sent to jail, there were no get out of jail free cards. Given this stated criteria, I nonetheless had a senior NCM approach me and tell me that he was the perfect fit for the job in Dubai. This individual had no prior experience in any welfare capacity, so I informed him that his name would not be forwarded for the job. The one thing that struck me as odd about his request was that he was holding a very senior position and he wanted to leave the operational theatre to run a leave centre. In the end, the individual in question was sent back to Canada as a stress casualty. This decision was not made lightly and I found the whole situation challenging. I had the choice of possibly helping this individual cover up his illness by sending him to Dubai or to stick with the prerequisites previously agreed to. Despite the challenges, I am nonetheless satisfied with my decision. Indeed, to this day, the individual in question has not returned to work. Furthermore, we did not have a single reported incident in Dubai.

Because of our state of operational readiness and the constant level of skills that we needed to keep up, the CO decided that we would train our soldiers at the Tarnac Farm facility. During one of our night training sessions, an American F-16 fighter aircraft dropped a 500 pound bomb on our soldiers. I felt the bomb blast in my tent six kilometres away. I definitely knew that something was wrong when the radio, which had routine range chatter on the net, went silent. I immediately started getting dressed to go to the command post (CP). The radio then started sending situation reports on what had happened. The final casualty count was four dead and eight wounded.

Immediately, another dilemma arose concerning the best spot for the CO to be at. Knowing that the situation was well in hand at the scene of the incident, I suggested the CO be with the wounded soldiers. This is where we were for the rest of the evening through to when the casualties were evacuated to Germany the next morning. Being at the CO's side throughout this challenging time became my number one priority.

A plan was put in place to ensure that the deceased were given a proper farewell on their departure from the Kandahar Airfield. A planeside service was deemed appropriate and, as many of the Battle Group that could be spared, attended the service. The most notable gesture made by our American comrades was that the Color Guards from each of the Brigade Combat Team Units were represented. My hat is off to the Brigade Combat Team Sergeant-Major who ensured that they were there for the service.

"A" Company, which was the company training at Tarnac Farms, was working up to go the Kowst airstrip on a company level security task that the Brigade Combat Team Commander had trusted us with. Kowst was a base camp for Special Forces and OGA. "A" Company was going to be based in Kowst for one month. Because of the Tarnac Farms incident, however, the mission was given to "B" Company. This proved to be a good decision as the Special Forces recognized former members of JTF 2 within the company and the transition went a lot smoother than if the unit would have remained an "unknown until proven." All this to say, because of previous Small Unit Exchanges, soldiers recognized each other on the battlefield. This in turn forged a bond that is unexplainable within this text. Suffice it to say, however, familiar faces go a long way on the field of battle.

After spending a week of mourning, the CO felt that we should go on an operation to get us out of the rut that we had entered. This resulted in another helicopter assault, this time in the Tora Bora region of Afghanistan. The mission was Sensitive Site Exploitation. The mountaintop that we were to assault was a defensive position used by the Taliban/ Al-Qaeda. The most important site exploitation that was carried out during this mission was on the cemetery where the DNA of Al-Qaeda members who had previously been killed in a laser bomb attack on their cave was collected. The names of all the soldiers involved in the cemetery operation were documented in case of post-traumatic stress. This cave site was also where Osama Bin Laden was last heard on his cell phone. We tried to move the roof of the cave to recover possible bodies. Unfortunately, however, this task was impossible with the tools that we had available to us. We then returned to Bagram and re-deployed to Kandahar with little or no problems.

One of the outstanding operations conducted by our Quick Reaction Force (QRF) at the Kandahar Airfield was when they were called out to go and secure an Apache crash site. The QRF was a platoon size task force that was on two hours notice to move anywhere in the theatre of operations. On this particular evening, the mission was to secure the crash site while the Explosive Ordinance Disposal (EOD) personnel would destroy all sensitive equipment that could not be removed. The mission started with the helicopters departing Kandahar to go to the crash site. When the helicopters landed at the Forward Armament and Refuel Point (FARP), the QRF thought that this was the crash site and dismounted. The helicopters took off and returned to Kandahar leaving the QRF at the FARP approximately 20 kilometres from the mission area. The Platoon Commander and Platoon Warrant Officer put their experience together and came up with the idea that the Spectra Gun Ship AC130 could give them the grid reference of the mission site. Notably, the QRF did not have a map of the mission area. All navigation would have to be done by Global Positioning System (GPS). They received the grid from the Spectra gun ship and proceeded on foot with the GPS to guide them to the site. All this was happening at night through unknown territory. These brave actions by the QRF again confirmed in my mind that all the harping that I had made about ensuring that the Battle Group understood that mission, buddy, self then everything else were the soldier's priorities in life had gotten through. The QRF made the Battle Group proud that night and their leaders were awarded for their contribution to the success of the mission.

Notably, an organizational change occurred at this point in the tour. The officer commanding (OC) and CSM of the Administration Company returned to Canada for career progression reasons. Because of the organization that we had on the ground, which included a Forward Supply Group that looked a lot like an Administration Company, when the OC and CSM left the operational theatre, the two organizations merged. This suggested to me that we had had a lot of redundancies when it came to logistical support. The soldiers carried on in their daily routines as if no one had left.

Operation Cherokee Sky was our last mission in Afghanistan. It was unique in the fact that there was a ground Assault Group and an Air Assault Group. The Brigade Combat Team Commander let the CO plan the complete mission as most of the soldiers that were involved in the mission were from the Battle Group. Besides the American assets that were required for the mission, there was an American Special Forces team that was attached to us to act as the liaison with the local Afghanistan forces. This was the first time that the "Coyotes," Light Armoured Vehicles (LAV), were used in an offensive role.

The mission occurred in Kabul province, approximately 100 kilometres from the city of Kandahar. A murder by the Taliban in a small town south of Qualat had instigated the mission. The Brigade Combat Team Commander wanted to prove to the Taliban and Al-Qaeda that he could project any type of forces anywhere in Afghanistan. The Canadian Battle Group, with the Recce Squadron, suited this mission. A forward operating base was selected near the Governor of the province's family residence. From this forward operating base a ground and air assault mission was launched on the town where the murder had taken place. Of course, by the time we had arrived at the town's perimeter and checked with the elders of the town, the perpetrators had disappeared. Nonetheless, the mission was well executed and all returned to Kandahar Airfield without incident.

CEASE OPERATIONS

The message to cease operations was given shortly after our return from Operation Cherokee Sky and all the equipment was prepared for pack-up. It was recommended that a pack-up team from Canada come to Afghanistan to carry out this task. This proposition was accepted by National Defence Headquarters (NDHQ) and a team was sent to Diego Garcia. This location, however, did not make any sense to me as they would have been able to do their jobs much better from Kandahar.

DIEGO GARCIA

The team stayed in Diego Garcia and we, as the Battle Group, brought everything with us from Kandahar. Once in Diego Garcia, the soldiers turned in all their mission essential kit and prepared for the next leg to Guam. Leaving behind mission essential kit was a good idea because it allowed for one-of-a-kind pieces of equipment that were procured for the Battle Group before the deployment and while we were in Kandahar to be well tracked. The items of particular note were all the C8 carbines, PAC4s, PAC5s and the NVG that had been collected from across Canada to outfit the Battle Group. The layover in Diego Garcia was between four and thirteen hours. The nicest thing about the layover was being able to have a shower and to change into civilian clothing for the trip to Guam.

GUAM

Spending a few days in Guam after the mission was something that the Commander and I had discussed prior to our deployment to Afghanistan. We both realized that we were going to change while on operations and the excitement of being immediately reunited with our families at the end of the mission might not be the best solution. The CO and I had both experienced coming back from a high intensity mission and simply going home and expecting to fit in as if we had not been gone for six months. Because of these experiences, which turned out to be filled with unrealistic expectations, and the anticipated operational tempo, it was decided that the Battle Group would experience some sort of re-integration period before returning to our families in Canada. One scenario that we bashed about was to take a cruise ship back to Canada from some European Port, then to board a train to our home bases. In the end, we settled on Guam.

Most in the Battle Group realized that it would be important to re-integrate rather than just do the standard thirteen hours of going from "shit hole to bed sheets." One thing to remember about having a re-integration policy is that in order for it to work, you must have a good screening process. This screening process provides the baseline to any measurement that has to be completed on the soldier during and after his deployment.

The whole Battle Group passed through Guam. Both the CO and I thought that this part of the operation was going to be just as important as anything else we had done to date. The soldiers had changed since their arrival in Kandahar and most of them had not noticed it. Some of the soldiers, especially the older ones, felt that Guam was a waste of time. However, it was good for the soldiers, particularly for the young ones, to attend the briefings. I would argue that the Guam experience was good for all of us. The interaction of the sub-units for all the activities including time-off was an asset to the reintegration process. The stay in Guam lasted at most three days for the soldiers.

EDMONTON

The return home to Edmonton was marked by a reception that the soldiers in the Battle Group will never forget. The streets were lined with people from the International Airport all the way to CFB Edmonton. The bus route took us through the main street of Edmonton and it seemed that the whole city had come to see us return home. For most of the soldiers in the Battle Group, it was the first time that they had seen the public react this way since they had joined the CF. Even with twenty-years of service, it was certainly the first time for me.

The half-days of work that we put in for the two weeks prior to the Battle Group going on leave was a good idea. Most of the administrative appointments could be conducted in the mornings, letting the soldier go home in the afternoon to have some well deserved quality time with their families.

Prior to us going on leave there was a welcome home parade in downtown Edmonton. The parade was another unbelievable experience. Even though it was during the work week, many citizens were in attendance. The newspapers later reported that there were between ten and fifteen thousand people lining the streets. This was quite the experience for the soldiers of the Battle Group. Our previous returns from peacekeeping missions had barely warranted a mention in the press. This remarkable reception indicated that the population had been informed, and approved, of what we had done. I credit this level of awareness to the CO for making the decision to bring the Canadian Press reporter with us and to have him embedded in the Battle Group from our first operation through to our final departure from Kandahar.

CONCLUDING REMARKS

Life after such a mission is hard to adjust to for some soldiers. They feel that they have served their country to their fullest and that they should be recognized for this sacrifice on a daily basis by being told repeatedly what great heroes they are. Most soldiers, however, live the life of a typical Canadian. They are very modest about their experiences abroad. They sit in the legion and converse with each other in a normal tone. If you ever want to find out about what really happened in any particular incident that the CF has been involved in from a grass roots level, then seek out these quiet soldiers who sit away from their boisterous counterparts. They are the ones who have the good stories for you.

CHAPTER 4

LEADERSHIP LESSONS LEARNED

Alex Bouzane

In order to underscore the leadership lessons that I learned from my deployment to South West Asia (SWA), I decided, after careful consideration and advice, that the best manner to do so is to compare operational experiences. Specifically, my experience on Task Force Aviano, on Operation Echo where I served as the Task Force Chief Warrant Officer (CWO) from September 1999 to May 2000, with my role as TAL Det CWO from 12 July 2002 to 14 January 2003 on Operation Apollo on the Joint Task Force South West Asia

Operation Echo was based in Europe and its mission was to enforce the no-fly zone over Bosnia. The detachment was composed of: six CF-18 Hornet aircraft; eleven pilots, of which two were the Detachment Commander and Deputy Detachment Commanding Officer; an Intelligence Section; an Administrative Section; a Logistical Section that included supply, transport and eventually cooks; a Military Police Section; a Public Affairs Officer; a Chaplain; and an Aircraft Maintenance Organization. Headquarters was based approximately 150 kilometres from the detachment with the Task Force Commanding Officer and staff. Although I was located with the detachment, I was responsible to the Task Force Commander.

Operation Echo consisted of approximately 110 personnel. Unlike Operation Apollo, I was deployed with my peacetime squadron. However, there were still many additional members assigned to the mission from Trenton, Winnipeg, Comox and Greenwood. All pre-deployment preparations and courses were held at our home base of Cold Lake, Alberta.

The SWA mission, Operation Apollo, as I understood it, was to deploy with a fully operational transport unit, which consisted of flight crews, a maintenance organization, an air movements section, a weather section, aircraft security officers, an intelligence section, which served two organizations, and a detachment orderly room. The National Support Unit was also included in this organizational structure and was responsible for providing support to all the operational units in theatre, including medical, religious, welfare, administrative, policing and logistical requirements. Additionally, the Long Range Patrol Detachment Unit, flying the CP-140 Aurora aircraft, carried out surveillance duties and the National Command and Control Information Systems' unit, a communication squadron, was responsible for all the secure and non-secure communication networks.

I was assigned to the SWA mission through 4 Wing, Cold Lake from 1 Canadian Air Division. Once contacted by the tasking cell, I committed to the mission request and started my deployment clearance, better known as Deployment Assistance Group (DAG), now known as the Annual Personnel Readiness Verification (APRV). Originally, I was not going to deploy until late August/early September, however, I was soon advised to be ready one week ahead of schedule for my one week mission course held in Trenton. I would be deploying immediately afterwards.

The first time that I met the TAL Det Commanding Officer (CO) for this mission was on the pre-mission deployment phase just prior to departure. I met the Deputy Commanding Officer (DCO) on the aircraft enroute to the mission destination. This was also the first time that I met the advance party members of the unit, some of whom, however, I had known prior to the mission. I would meet all the TAL Det Roto 1 members over the next two weeks as they arrived in theatre.

As I came to the Detachment CWO position from outside of the mobility community, I was lacking the one sure way to get to know the team; I had not previously worked with them. I discussed this difficiency with my CO and had his full support to pitch in wherever and whenever I felt it would be beneficial. I worked building up pallets with the Air Movement personnel, made daily visits to the Metrological and Intelligence sections and I spent time on the aircraft ramp with the maintainers. I also flew with each crew on operational missions. I was impressed with not only the professionalism of the entire crew, but also the C-130 Hercules tactical arrivals and departures. During the change over and subsequent arrival of all members, I spent a great deal of time visiting with all sections in the unit and much time with the CO and DCO in order to assure that we were on the same wavelength.

The Tactical Airlift Mission was to provide Tactical and Strategic airlift capability to coalition forces serving in SWA or, as our CO would say, "we deliver trash anywhere/anytime 24/7." Throughout the tour, the CO and I would visit with the troops officially and unofficially to brief the mission, discuss day to day situations and how they affected operations, and to receive feedback and provide support.

TAL Det worked twenty-four hours a day, seven days a week. Once a week an "A" staff meeting was held to address issues, as well as requirements for ongoing missions. This provided all section heads with up-to-date information for planning work requirements. Additional meetings were held as required.

The TAL Det Team of approximately 140 personnel on Operation Apollo was composed of personnel from across Canada that had begun team building on day one. It was composed of flight crews and Air Movements personnel from Trenton and Winnipeg, aircraft maintenance technicians and operation room (OR) staff from Trenton and Winnipeg, aircraft security officers, and weather and intelligence personnel from Trenton, Cold Lake and Bagotville. As previously noted, the unit was assigned to a camp along with a National Command, Control and Information System (NCCIS) unit and another Canadian (Long Range Patrol) as well as a foreign operational flying unit. The personnel assigned to the camp as support to the operational units had a different operational chain of command.

There were some major differences between the two operations. Operation Echo was primarily one team working together towards one goal. Operation Echo had one commander for both operations and logistics. Conversely, Operation Apollo had one commander for each, a tactical operations commander and a camp commander for logistical requirements, with the latter not being in the chain of command of the flying units or involved with daily flying operations. There is no doubt that this multiple command structure caused friction.

There were in fact many internal and external factors that would affect the "Team" concept during Operation Apollo, some positively and some negatively. TAL Det was the only unit with a twenty-four hour a day, seven day a week operational status. Moreover, the unit was required to meet real life commitments and operate under severe hot weather conditions.

These challenges became a rallying point for all, as this created an extreme sense of pride in the accomplishments of the unit. All members were consistently reminded and congratulated on this outstanding accomplishment. The need to be identified as a unit came to the forefront when a suggestion was made to have all in theatre wear the same dress. This was a significant issue and was met with great resistance from the different units. Immediately, a decision was taken to clearly identify the TAL Det members by wearing a subdued TAL Det crest on their daily work dress. This amendment proved extremely popular. The need to be identified as a unit within the larger group at the camp was not only necessary for morale, but also for team identification. All units adopted a clear, albeit subdued, unit designator.

During Operation Apollo, different units were initially housed together. This combination proved to be not only a morale issue, it also quickly became an operational one. Subsequently, units were housed in close proximity to each other, however clear unit lines were established. Although some may consider this small and, possibly, insignificant the importance of unit lines and the pride put into their upkeep was evident in the camp. The unit lines proved to be a gathering point for all personnel, not only TAL Det. Many innovative items and unit identifiers were clearly displayed to mark lines.

As with all organizations, however, there were non-team players and, unfortunately, when such a member held a supervisory position, this caused conflict with peers and difficulty for subordinates. In one case, when all reasoning and mentoring failed, a direct approach had to be taken and the member was counselled. On other occasions, where conflict or rivalry existed because members were from different units prior to joining TAL Det, something as simple as the unit soccer team proved to be an excellent team builder. All members, regardless of whom they worked for before, came together to show solidarity and teamwork to all the other units.

The Christmas dinner of 2002 highlighted the importance of teams. As usual, TAL Det flew every day, even on 25 December. Nonetheless, the rest of the camp decided to dine at 1700 hours (hrs). All of TAL Det, remained on duty until the last flight was down and serviced in order to have Christmas dinner together.

One of the most difficult tasks on operations is to maintain constant vigilance regarding mission importance. Unfortunately, during periods of high activity and stress, especially when extended over a long period, this requirement is severely challenged. Constant briefings are not the answer unless you can provide concrete information to substantiate and explain the situation. Focusing on intelligence reports to identify the danger only works when there is factual information to brief; otherwise, the troops take it as just another report.

As good intentioned as the changes may be, without proof to substantiate it, the troops see it as a make work project. These perceptions detrimentally affect the troops morale. An example of well intentioned changes to the normal routine that ended up frustrating troops on Operation Apollo was when the drop off location for members of the camp who were permitted travel into a nearby city was changed. Originally they were dropped off at a very busy location where they immediately mingled with tourists and western locals. Later, personnel were being dropped off and picked up at locations that were obscure and the only people that were there were the troops. For the perceived sake of safety, they were actually now more vulnerable than before. Regardless of the observations about this issue being put forward by the troops, leadership did little to reverse the well intentioned change. However, when real intelligence was provided in another scenario, the situation was easily addressed and the restrictions on travel were readily accepted.

Due to the wide range of rank and professions of the personnel on the detachment, including a large number of aircrew, Warrant Officers, and senior non-commissioned members (NCMs), solid, factual and believable information was the key to a successful briefing. It is difficult to provide veiled intelligence to experienced NCMs and Warrant Officers, let alone senior officers and flight crews who have flown into combat zones. These individuals demand full support and credibility from those around them.

On this regard, Operation Echo provided a different set of challenges. Fighter pilots, except for the Detachment Commanding Officer and DCOs who deployed for a full three months, were changed out every thirty days to maintain flying qualifications. All the remaining members, however, completed a full six month tour. Nonetheless, as we were one camp, the team came together readily.

One advantage was our location in Europe, which provided excellent opportunities for improving morale. The only real morale challenge was the poor quality of the mess which was available through the base that we were housed at. Due to very limited hours of operation and food choices, many members chose to eat elsewhere. This proved to be quite expensive, as the members were only provided mess hall per diem. Consequently, a great deal of time and energy was spent convincing command that a dining facility was required. Finally, after three months and a visit by a senior commander, who joined us for several meals at the local dining facility and was served the same meal each time, the point somehow managed to get accross to those in charge. With the commander's backing, a mess facility was erected after Christmas. In the meantime, two memorable occasions were Thanksgiving and Christmas 1999. As we had no cooking facilities on site, we borrowed the use of a kitchen and the officers and senior NCMs prepared and served a complete traditional dinner to all the troops.

During Operation Echo, personnel were housed in trailers similar to those we would later use during Operation Apollo. Since we were only one unit in 1999/2000, there were no real issues. Flying officers were provided single billets because of their constant rotation and small number.

From the outset, mission focus was apparent on Operation Echo. Each day, fully armed aircraft were launched and the ground crew waited for their return. The pilots shared their mission debriefs with the maintenance crews and constant intelligence briefings were given to keep everyone in the loop. The local area was a breeding ground for the Red Brigade, which was very active. Although the main focus of their belligerence was our coalition members, as we were based together, the same security vigilance was required for everyone.

The Operation Echo mission never varied throughout the tour. Climate and location were non-factors. The TAL Det Operation Apollo mission requirements were also relatively stable throughout the tour, with the addition of the occasional humanitarian relief flight. However, in SWA the weather was a real issue. Nonetheless, from the most junior member on the ramp to the most senior flying officer, the importance of delivering the package to its destination was always forefront. Even under the most trying conditions of 50 degrees Celsius weather on an open ramp – when the touch of the aircraft would burn your skin – major aircraft maintenance was accomplished in exceptional time. Flight crews spent gruelling hours flying to isolated fields and without shutting down, under armed guard, unloaded very valuable cargo. These were always high-tension missions that exacted a cost on all involved. Nevertheless, the importance of our missions was seen by all, even those who did not fly. At times, special teams were sent into locations to repair aircraft or prepare loads for the aircraft. When the location was remote, or an area of possible high threat activity, the personnel were always extremely professional and demonstrated excitement with the opportunity of completing their job as they were trained to do.

As with any mission, however, the closer to the end, the more people became distracted. At this time, mission focus became the topic of every briefing. The occasional detachment social was instrumental to good morale. A simple day playing sports or taking advantage of a "Camp Welfare" tour went a long way to breaking up the daily work routine and provided the opportunity for members to recharge their energy levels.

As with all missions, on Operation Apollo there were personnel with personal problems or who had family emergencies. When that person is senior in rank, then it affects all those below him or her. The importance of good screening not only by the social worker, but also by the member's supervisor cannot be stressed enough. One such individual required continuous counselling and mentoring. Moreover, this individual could be extremely difficult to deal with, especially when the member would show uncontrolled emotions at work. Eventually, counselling for personal problems turned into counselling for poor work performance and unprofessional behaviour. We nursed this individual through the tour because I believed that returning the member early would have done more damage to the member. With the exceptionally strong Warrant Officer cadre, this was eventually handled within the detachment, although it generated much conversation with the CO and DCO.

On both missions, many individuals encountered personal problems, from marriage disputes to legal issues. All supervisors should be aware that many members try to hide these problems prior to deployment. A good, thorough interview prior to deployment could likely catch major issues beforehand and limit the need to repatriate personnel early. Certainly, individuals with problems leading up to deployment still have them when they deploy. Moreover, they are often amplified by the stress of separation.

This was the case in several situations that I encountered. On one such occasion, a member was facing legal issues, which ended up resulting in military charges. Although the unit forwarded the charges and expected the trial to be carried out in theatre, this was un-realistic, as no evidence, reports or witnesses were available. The report was returned and the unit was advised to proceed when the member returned.

Additionally, during operations, the padre's office was well used by several individuals, as was the medical office. Most of the time, a willingness to listen, provide advice and then follow up, solved many situations.

Two items that were considered major morale boosters for each mission and that provided an opportunity for the members to recharge was the Rest & Relaxation (R&R) and Home Leave Travel Allowance (HLTA). These are excellent programmes that were used to the fullest during both operations.

Notably, an interesting part of any deployment is the "Code of Military Discipline" and how it is applied. Most importantly, troops want to know that it is being applied equally across the board. Troops on both Operation Echo and Operation Apollo were affected by this perception.

In theatre, breaches in discipline will most likely be committed by those ready to repatriate or by those just arriving. For both missions, the beginning and end of the tour was the time when the most breaches of discipline occurred – despite a two week compulsory "dry period" at the beginning and end of Operation Echo. For example, during Operation Apollo, shortly prior to returning home, two senior personnel broke curfew. This situation was observed by not only those in TAL Detachment, but by everyone in the camp. A military police investigation was conducted and charges were laid and eventually disciplinary action was taken at home bases.

Unlike the land and maritime elements, air operations are not familiar or comfortable with disciplinary situations involving senior personnel. However, I believe this will become resolved with the increasing tempo of operations. In another disciplinary situation in the Camp, which involved an NCM from TAL Det and a senior member from another organization, charges were also laid against the senior member. Thus, justice under the Code of Military Discipline was seen as fair and equitable.

When disciplinary rules are not applied evenly, not only does morale suffer, respect for operations and supervisors also quickly diminishes. Despite even the most valiant efforts of senior NCMs and Warrant Officers, if it is not perceived to be fair by the junior members, then no amount of justification or explaining will get them on board. In theatre, all members have to be held accountable to the same rules. Although this was uncomfortable at times, the end results were always just. This is not to say all breaches to the rules require disciplinary action; rather, it means that everyone is treated the same regardless of rank.

No matter where the operation is being conducted, there is a certain amount of increased risk to individuals. The risk level for flying over uncontrolled land or airspace is generally higher than that at the base camp. The risk assessment at Camp Mirage, SWA, was normally low to medium, yet, based on situational intelligence reports, it could periodically be increased. When that happened, force protection was increased. The danger of overreacting to this type of scenario is that the importance of force protection is lost when it is really required, as the personnel will not perceive it to be any different than normal, akin to the classic tale of the boy who cried wolf.

Moreover, when senior staff are stressed and concerned, it is very hard to convince junior personnel that all is well. While there were situations that caused anxiety amongst some personnel, the discussion and response to those situations by senior and experienced members during Operation Apollo proved invaluable. Additionally, the sharing of information and operational focus of the crews was truly inspirational. When with them, even during higher risk missions, one felt completely safe. Jocularity, bravado and team work were signs of heightened anxiety and stress. Talking about missions and discussing possible scenarios assisted members in dealing with the situation.

Stress was most apparent near the end of the tour. At the beginning, everyone was anxious to get there and get the job done. Early on in the tour, very little time was consumed on what might happen or go wrong. As the end of the tour neared, however, the possibilities of personal harm or problems were underscored. This heightened anxiety affected everyone, not just the flight crews. Nonetheless, on Operations Echo and Apollo, I believe that every member of the team demonstrated a strong desire to participate and do a good job, as well as take responsibility for the success of the mission.

One of the greatest fears of any CO, or any supervisor for that matter, is the loss of someone while on duty, be it at home or abroad. Prior to deploying leaders bravely and confidently tell their superiors that they will return with all personnel. This responsibility is omnipresent. Even the thought of a careless accident causing the loss of a team member is always on your mind. I think that personnel in command and supervisors experience this anxiety more than others. It is not something you can dwell on, however. Rather, you must use it as a check and balance measure when making decisions regarding everything from high-risk adventures to normal operations and even R&R periods.

Both missions placed the flying crews in more intimate danger than the ground crews. Operation Echo saw only the pilots fly over hostile area whereas Operation Apollo at times had ground crew on board for mission requirements and military and even sometimes civilian passengers.

During Operation Apollo the stress of the mission and personal fatigue affected most personnel, regardless of rank or occupation. The climate was extremely hot, which took a measure from each individual. This was compounded by the effects of a change in diet. Within the first two months, the majority of the unit had suffered through a bout of jet lag, heat exhaustion and sicknesses associated with the new and different food and location. These strains manifested themselves in the work being conducted.

The main concern with sickness, however, is that it can easily bring down a complete flight crew; therefore great consideration was given to housing in order to prevent the passing of sicknesses from one individual to another. This ended up being an issue that caused lengthy discussions. The intent of the camp administration was to double up as many people as possible in order to have beds readily available for visitors and staff inspection visits. The TAL Det concern was the loss of a serviceable flight crew due to sickness and the change around of crews to meet mission requirement. Flight crews are fixed crews, no different than a tank or gun crew; they team up to be more efficient and therefore can handle contingencies and emergencies much better than a crew that is just placed together for a single mission. With limited flight crews, one down impacts on the remainder and careful monitoring has to be done to avoid over flying the flight crews without proper rest. These were constant concerns during Operation Apollo, however, we were ultimately able to defend the flight crew policy and members had the appropriate amount of rest.

In general, during Operation Apollo, most of the team remained healthy. Physical fitness programmes were encouraged and the majority of the flight members engaged in a regular exercise programme. Not surprisingly, there was a keen interest in the soccer league against the other units. This was certainly a very healthy programme where all units became involved, either through playing or just showing up to cheer on their team. All exhibited a great sense of pride. The importance for stress relief of participation in a sports programme cannot be overrated.

During Operation Echo, as well, the complete unit got behind and supported the soccer team as we played against the Spanish team. Being completely outmatched was of little concern. At the time, we had a senior officer visiting the camp and he put on his soccer gear and played valiantly for our team. This one act clearly stated the concept and meaning of being part of the team. Although we lost by a wide margin, we unquestionably won the gold for team play. The number of people limping around camp over the next few days demonstrated the effort and spirit put into the match.

Visitors to the camp had many different effects on the personnel. I will endeavour to write the feelings and impressions that were expressed to me in regard to those who visited the camp during my Operation Apollo tour without being overly critical.

One of the highest profile visitors was the Governor General. The entire camp was excited about her visit. She and her husband endeared themselves to all and took the time to ensure that they visited with or, at least spoke to, every single person. The Governor General was truly seen as our Commander-in-Chief. Her message and empathy were felt and heard every time she addressed the camp. She had a tremendous positive impact on all of us. However, this boost to morale did not occur with each visitor.

Many visitors, once they did their required visit to the camp, quickly ventured with an escort party to see the local sights and visit the bazaars.

The purpose of their visits was seen as a shopping opportunity more than a goodwill visit to commend serving members. As a CWO, I found their behaviour extremely hard to defend in front of the troops and most of the time brushed it off as just part of the visit agenda. It appeared that considerable time was put into the agendas of visiting ranked personnel, officers and NCMs alike, with a very limited amount of that time being spent at the camp, however. Although we had a very good mess hall, which the members dined at daily, there seemed to be little interest from senior military visitors to join us for a meal.

This apparent lack of interest by visitors to the camp was not as obvious during Operation Echo as it was for Operation Apollo. In regards to Operation Echo, many high profile senior officers and very important persons (VIPs) visited the mission. Amongst these were two honourary colonels. These two individuals were exceptional ambassadors for both the CF and the people of Canada. Throughout their visit they constantly assured all members that what we were doing was for the greater good of humanity and that we had the full support of the Canadian people. They spent their entire time in camp, listening to the troops. Although hotels in the local area or VIP quarters were available, they stayed in the trailers, amongst the troops. It is most unfortunate that due to the political atmosphere, security concerns and the location of the camp, during Operation Apollo, honourary colonels were not authorized to visit. These visitors would have received a warm welcome and their visits would have been eagerly anticipated. Furthermore, the benefits of their visits would have been apparent instantaneously.

I would be remiss if I did not mention the Canadian Forces Personnel Support Agency (CFPSA) Show Tour and highlight the exceptional work that they do for deployed military members. The CFPSA Show Tour was without a doubt a major crowd pleaser and, certainly, an outstanding morale booster. The members of the Show Tour were exceptional in their ability to mingle with the troops and officers and to show their appreciation for serving CF members. They were true ambassadors of Canada. At Christmas, every single individual at the camp received a large box full of presents from CFPSA. I cannot say enough or give enough praise to the members of the CFPSA staff who put this together. These acts were very much appreciated. Importantly, the Militray Family Resource Centre (MFRC) provided another medium for the families to correspond with those who were serving.

Notably, the cultures of the host nations were quite different on Operation Echo and Operation Apollo. Operation Echo was based in Europe and it was easy to assimilate to the local culture as well to our hosts. We had two host countries and both were very much inline with Canadian values. In contrast, Operation Apollo was in SWA and a great deal of time was spent on briefing our personnel on local customs and the religious atmosphere. Great care was exercised by all supervisors to lead by example and to constantly remind our personnel, especially during religious holidays or special cultural times, to be aware of local customs and to be respectful of local religious values. In particular, our female personnel needed to be aware of their situation in the local environment and to dress conservatively by our standards in order to avoid unintentionally offending anyone or drawing undue attention. Moreover, Canada's openness to same sex relations has not necessarily been accepted in all areas of the world. This is without a doubt an extremely sensitive issue to address.

Of course, during both operations sexual relations and fraternization were openly discouraged – and the CF made it clear that they were forbidden on operations – however, intimate relations were extremely hard to police. The selection of senior NCMs and Warrant Officers of both sexes is vital to provide council and guidance to members in order to minimize problems in this area. Officers and NCMs are alike in this issue and fraternization, especially between senior and junior members, can be demoralizing to the entire camp. This issue must be emphasized for the senior personnel especially. The importance of setting the correct example is paramount to success in this area. Moreover, if the issue arises, it must be dealt with immediately, even if only to let the individuals know that there is a perception that something appears amiss as perceptions can speak as loud as deeds in the closed environment of a deployed camp. On two occasions, the members did not realize the impression they were making and, once informed, took corrective action.

Technology on each of these operations was the key to success. The ability for a technician to call an aircraft representative or home base in regards to an unusual technical problem prevents lengthy delays in that operational asset becoming serviceable again. The emergency phone call home because a family member is sick, provides you with an airman, soldier or sailor who has their mind back on the mission. The importance of keeping people informed not only of what is happening on site, but the news from home, is vital.

The set up on the Defence Wide Area Network (DWAN) on operation is the difference between darkness and daylight. It is essential to today's operations. The number of problems solved, situations sorted out and advice sought through this network is uncountable. DWAN was the lifeline on many occasions and without it we could have lost valuable assets.

A major morale benefit was the ability for the members to have access to e-mail and a once a week telephone call. For members this was their instant contact with the home front and it provided an opportunity to share a moment with their spouse or children. Regardless of all of the other technological resources, DWAN was the network that helped many retain bonds to loved ones at home. From a military point, the access to the DWAN and Central System Network (CSN) system are immeasurable.

Overall, during each tour, our successes greatly outnumbered any negative points that may have occured. One particular occasion from each tour underscores the positive atmosphere that existed on each operation. During Operation Echo, prior to the construction of the mess tent, as I previously noted, the officers and senior NCOs joined in cooking and preparing Thanksgiving dinner for the detachment. This was a challenge, however the complete meal was prepared at a satellite kitchen and transported to a large tent that was used by the detachment as well as others. A complete turkey meal with all the trimmings was served cafeteria style. The whole detachment sat down together and enjoyed dinner as a large family would have at home. As this was early in the tour, it demonstrated what a little planning and hard work could achieve and set the tone for the remainder of the mission. The benchmark occasion during Operation Apollo was the detachment medals parade. To demonstrate our uniqueness, with the CO's approval, I planned the parade with my senior NCOs and warrant officers. The detachment formed up in the shape of a C130 Hercules aircraft. Flight crews and Headquarters staff were the engines and cockpit with the crews making up the wings, empennage and tail. Personnel were presented their medals on site with the remainder of the aircraft standing at ease. This was indeed a very different parade, however, as important as any parade and carried out with the same respect and with the appropriate military flair. There is no doubt that every member of the Tactical Airlift Detachment Rotation 1 will remember when they were awarded their medal.

Each mission presented unique challenges, which in turn provided diverse opportunities for individuals to demonstrate their skills and abilities. If I had to choose a single issue we should focus on improving, it would be in selecting and approving personnel for deployment. Supervisors must be held accountable when signing a member off for a tour. They must be accountable to their CO, assuring the CO that the member is prepared for the mission and indeed has no family problems at home. A simple interview with a CF social worker is insufficient as members can easily provide a story that works for, or against, joining the mission. Personnel who fail the APRV or who fail to provide the complete story and are required to be repatriated should be held administratively accountable. Moreover, the Detachment CO should personally interview all senior personnel, at least when practicable. If assigned outside of the mission, at least he or she should be contacted by the member's CO and have a telephone interview with the mission CO.

In summary, the leadership at all levels during both tours was of a very high professional calibre. Even in times of increased stress and heightened security, the leaders remained calm and were consequently effective.

CHAPTER 5

"LOOSE LIPS SINK SHIPS": THE LEADERSHIP EXPERIENCES OF CPO1 KEN MULLINS ON OPERATION APOLLO, ROTATION 2

Craig Leslie Mantle[1]

Between September 2002 and April 2003, Her Majesty's Canadian Ship (HMCS) *Montreal* deployed to the region encompassing the Gulf of Oman and the Persian Gulf on Rotation (Roto) 2 of Operation (Op) Apollo, Canada's contribution to the US-led War on Terror.[2] Over the course of their deployment, the ship's company conducted numerous hailings and boardings of foreign vessels in an attempt to restrict the flow of Al-Qaeda and Taliban terrorists and their equipment, be it weapons or otherwise.[3] When not participating in such activities, they escorted allied vessels, usually high-value targets, through the Strait of Hormuz, a narrow passage that connects the two gulfs. All in all, they affirmed Canada's (and the coalition's) naval presence in the region and, to a great extent, controlled the movement of both people and materiel on the water. Throughout this trying period, and even during the lead-up to the deployment itself, Chief Petty Officer First Class (CPO1) Ken Mullins, *Montreal*'s coxswain, faced a number of leadership challenges that forced him to rely on his experience, judgement and imagination to solve.

In order to relieve those ships already deployed on Op Apollo, HMCS *Montreal* was scheduled to sail from Halifax in September 2002. While working in Ottawa in January of that year, CPO1 Mullins received a call from *Montreal*'s current coxswain asking him if he would be willing to take over in December, midway through the deployment; the latter was seeking a replacement as his current posting as the most senior Non-Commissioned Member (NCM) aboard ship was scheduled to terminate at year's end. After taking some time to consider the offer – the prospect of leaving his desk job behind in order to go to sea had already piqued his interest – CPO1 Mullins decided that it would be best, for both himself and the ship's company as a whole, if he had the opportunity to participate in the pre-deployment exercises, or work-ups, rather than join the ship after these critical trials had been completed. After numerous phone calls and e-mails to both the current coxswain and *Montreal*'s

Executive Officer (XO), it was eventually decided that Mullins would participate in the entire deployment cycle from start to finish by joining the ship in Halifax in May.

Leaving Ottawa prematurely allowed him to contribute to work-ups, a critical period in which the ship and its crew were tested for operational readiness. During this time, he set the tone for his term as coxswain by making both his intentions and leadership style known. Aside from demonstrating his abilities (and thus his competency) to both his subordinates and superiors alike, he observed how the ship's company operated and corrected mistakes where possible. At this point as well, he expended considerable energy on team building exercises and endeavoured to develop trust between all concerned. Had he joined later in the deployment, thus disrupting the ship's already-established rhythm, he would have been forced to create trust and to team build under very demanding circumstances that allowed little, if any, margins for error. The fact that he had not "been there" during work-ups would also have hindered his progress, as he would have joined the ship after much of the hard preparatory work had been completed, thus depriving him of the shared experiences so essential to the growth of emotive bonds between personnel. The chances that he would have been viewed as an outsider, at least until sufficient time had elapsed for him to be fully integrated into the ship's company, was a risk that he was not prepared to take and thus he attempted to avoid this situation altogether.

CPO1 Mullins faced a number of leadership challenges even before leaving Canada. He had known from the outset that HMCS *Montreal* was to be deployed on Op Apollo as a command platform, in that Commodore Dan Murphy, a Canadian, would command a number of multi-national coalition vessels from this ship. To this end, much of the available space was converted into offices and other facilities to house the apparatus of command, such as additional communication systems. Although not originally designed to accommodate a command staff, the *Montreal* was adequately converted to meet these new requirements. Originally, there had been no intention of taking a helicopter detachment along, however when the *Montreal* returned to Halifax from work-ups, it was decided that one was required. Indeed, the presence of a Sea King would provide added operational flexibility. In addition to the berths required by the commodore and his staff, space for even more bunks now had to be found to accommodate the air force personnel who would fly and maintain the helicopter. Anticipating the probable outcome of events, Mullins believed that he would eventually be told to leave certain members of the ship's normal complement behind in order to house the added personnel. In his capacity as coxswain, though, he took it upon himself to advise his superiors that such a course of action would be exceedingly detrimental if it ever came to fruition, for if the ship sailed with less hands, the remainder would be obligated to "pick up the slack," thus placing additional burdens on them. He predicted that an insufficient number of sailors could potentially cause difficulties later in the deployment and thus took immediate steps to ensure that such did not occur. In the end, the decision was made to sail the *Montreal* with its full complement.

Faced with the immediate requirement to find the space to house the extra berths, CPO1 Mullins and the XO toured the ship and identified those areas that they thought would be suitable in which to add additional bunks. A naval architect whom they later consulted determined that the selected spaces were not acceptable and so the process began anew. This time, space was created by removing the settees in the messes, thus depriving many of their quiet areas that they habitually used to relax and to reflect on both the day's events and tomorrow's challenges. Given that the *Montreal* was scheduled to depart Halifax shortly, there simply was not enough time to consult the sailors about this drastic, yet necessary, change. The master-seamen and below were naturally quite upset about the new arrangements as the reorganization had affected them considerably. When time later presented itself, CPO1 Mullins made a determined effort to explain the reasoning behind this confiscation of personal space to "anyone and everyone" who would listen.[4] In part, he explained that when the boarding parties were eventually sent afloat in theatre, which would be composed of *Montreal*'s own sailors, those who remained onboard would have to "fill-in" for these temporarily-absent crewmembers. When combined with the natural loss of some sailors, such as those who would be forced to return to Canada during the deployment for personal reasons, the remainder would have to face a much greater workload. If they were already many sailors short, the burdens of this additional work would eventually become extremely onerous. In contrast, so he explained, the presence of the regular number of sailors would alleviate much of the stress and strain of overwork that would surely occur if the ship carried fewer personnel. In fact, he had used many of the same arguments to persuade his superiors that a full complement was truly required.

Once the situation had been explained, the sailors onboard the *Montreal* agreed that the extra hands were necessary and that they would ultimately benefit from it, but they were less than enthusiastic about the new arrangements. The sailors' unhappiness was further muted by the fact that all aboard ship had suffered to one extent or another. The officers and petty officers (POs) also lost space in their respective messes in order to accommodate the extra hands. The burden did not fall entirely and exclusively on the lowest-ranked sailors alone, and when they learned that everyone was sharing in this sacrifice, the new arrangements became somewhat more palatable. More significantly, however, CPO1 Mullins allowed the now-displaced "rest" areas to be placed in certain "work" areas on the ship, thus providing the sailors with a new quiet place to relax. Although they would have preferred to stay in their own messes, to which they had become accustomed, they at least had some space where they could pause. Allowing the sailors to rest in designated work areas was usually not allowed, but the ship usually did not carry extra crewmembers either, and considering the situation, it was the best solution that could be found with the time, space and resources on hand. Had this situation not been explained, had the space been found only on the sailors' mess decks and had not some attempt to provide them with quiet areas been made, resentment and frustration would certainly have run high which would have, in turn, negatively impacted both morale and efficiency. In Mullins' opinion, the sailors would have accepted these changes, owing to their professionalism and commitment to the mission, but certainly would not have been as content as they were once the requirement for the changes had been fully explained to them.

Other challenges also presented themselves during the course of the deployment that again affected the crew's morale. Nearing the end of its deployment, HMCS *Iroquois* and HMCS *Fredericton* were sent to replace the *Montreal* in theatre.[5] Just outside of Halifax, however, the helicopter aboard the *Iroquois* crashed back onto the deck of the ship causing damage to both. Fortunately, the accident only caused two non-life threatening injuries. As a result of this incident, the *Iroquois* returned to the dockyards where a determination would be made as to its seaworthiness and suitability to continue with the deployment.[6] The command team aboard the *Montreal* – the Commanding Officer (CO), the XO and the coxswain – learned of the crash almost immediately and were thus forced into the difficult position of deciding when the crew should be told as the incident might delay their return to Canada. For whatever reason, the CO decided that the crew would be told of the crash the following day. Again in his capacity as coxswain, CPO1 Mullins approached the CO and asked to speak with him in private. When alone, he strongly advised him that because the *Montreal* was still attached to the outside world through e-mail, telephones and television, it was really only a matter of time before the ship's company would learn of the crash. Learning that they would be delayed through such impersonal means, he reasoned, would not benefit morale and so he advised the CO that the crew should be told immediately. Mullins believed that all onboard would appreciate hearing such news from their leaders rather than through the media and contended that the sailors would surely question why they were not told before the rest of the Canadian population. In the end, the crew was informed of the accident immediately.

The decision whether or not to tell the ship's company of the crash was not the only challenge that CPO1 Mullins faced at this time. News of the *Iroquois*' accident travelled fast and with more information now at their disposal, those aboard the *Montreal* began to worry first about the safety of their fellow sailors and aviators and then about the potential delay in their return to Canada. With the crew already exhausted from a long deployment and wanting very much to go home, news of the crash caused morale to dip for they assumed, quite naturally of course, that their presence in theatre would be prolonged. Rumours started almost immediately that the *Montreal* would remain on station for longer than anticipated. "Estimates" on the delay varied from a few days to a month. Some gossip even had the ship remaining in theatre until June! These rumours had the usual effect of making matters seem worse than they really were.

Some crewmembers in leadership positions became so distraught with the prospect of being delayed that they began to further influence the morale of their sailing mates through their demeanour and conduct. One individual in particular, a PO, began complaining with great emotion to his subordinates about the ship's possible fate. Coming from their leader, whom they trusted and looked to for guidance, these sailors began to believe what was being told to them and they too became pessimistic about their future prospects. Realizing that this particular PO was causing unnecessary stress, CPO1 Mullins pulled him aside and, after indicating that his conduct was entirely inappropriate, related that the *Montreal*'s leadership, from the CO to the master-seamen, had to retain a positive outlook for the benefit and morale of all. CPO1 Mullins informed the PO that they were not officially delayed yet and that this truth should be communicated to everyone onboard. All that was known for sure was that an accident involving the *Iroquois* had occurred. These rumours had to stop immediately and that the leadership especially should not be propagating misinformation, but rather attempting to dispel such exaggerations. As it happened, the *Iroquois* suffered no major structural damage and, after an inspection determined that all remained serviceable, the ship sailed from Halifax after only a few days in port. Being slightly delayed by the crash, the Montreal spent Easter away from home, thus causing the crew to miss a second important holiday in the course of their deployment – they had already been at sea on Christmas. Fortunately, the ship was proceeding to Canada in the early spring and so the crew were generally willing to tolerate this additional period of separation for they realized that their time away was nearing an end.

Another rumour seriously impacted morale and forced CPO1 Mullins to act immediately in order to mitigate its negative effects. Early in the deployment, word reached the *Montreal* that the Navy was considering the possibility of leaving the ship in theatre once the rotation had been completed and simply replacing the present crew with fresh personnel. The sailors, who had naturally grown attached to the ship in much the same manner as soldiers grow attached to their parent regiments, were very much opposed to this possibility. They desired to serve as *Montreal*'s permanent company, rather than its transient crew. To counter the negative effects of this rumour, CPO1 Mullins made a concerted effort to remind everyone that it was simply that, a rumour, and that it should not be allowed to impact performance. The Navy, he counselled, was simply thinking of the idea and nothing more. Nevertheless, the effect on morale was significant. In response, CPO1 Mullins and his six departmental chiefs, all of whom held the rank of either Chief Petty Officer Second Class (CPO2) or Master Warrant Officer (MWO) as in the case of the air department, wrote a collective e-mail to the Command Chief Petty Officer (CCPO) in Ottawa that outlined their thoughts, feelings and concerns on the matter.[7] Their correspondence encouraged the CCPO to persuade his colleagues, which happened to include the Chief of the Maritime Staff and other high-ranking officers, to cease making their views public on this issue for such commentaries were reaching the ship through e-mail and causing considerable angst and dissatisfaction amongst the crew. In the end, such discussions stopped owing to the influence exerted by the

CCPO. Despite this aggressive, and perhaps unorthodox, approach, another ship's captain was at the same time telling his own crew that they would replace the sailors aboard the *Montreal* and that they should prepare in earnest for such an eventuality. Again, through the perils of e-mail, these messages eventually reached those aboard ship. When, however, discussion on this matter stopped at higher levels (or at least was kept more confidential), the rumour slowly faded away and morale consequently improved once it was realized that in all probability they and the ship would return to Halifax as one.

Dispelling persistent rumours was certainly not CPO1 Mullins' only task during the deployment. All in all, he tried very hard to form a cohesive team with his subordinate departmental chiefs that would function well in all situations. He reasoned that those aboard ship would surely notice if one or two members of this small and intimate group failed to work well with others, a situation that might lead the more junior sailors to question their leadership's competence and skills, especially if tension and animosity accompanied such interactions. In CPO1 Mullins' opinion, presenting a united front in which all worked together effectively would encourage their subordinates to place their trust and confidence in them. He very much wanted to have his team "on the same page."[8] For this and other reasons, he insisted that he be present for *Montreal*'s work-ups, as this period would allow him and his chiefs to begin to form the bonds so essential to working together as a cohesive whole. He also made it clear at this point that he was willing to listen to them and that, if the situation so warranted, he would take their legitimate concerns to the CO on their behalf. During the deployment itself, CPO1 Mullins saw the various social events held aboard ship as an opportunity to team-build for, in his opinion, those who played well together tended to work well together. As a result, he strongly encouraged his chiefs to attend such functions, although the one or two individuals who were not so inclined were not pushed into participating since he realized that forced involvement would only lead to resentment and frustration. His efforts to form a cohesive team seem to have succeeded. On one particular occasion, for example, a sailor remarked, perhaps a little too loud, "Here comes the coxswain and his posse."[9] Rather than becoming angry – the statement was intended as a minor slight – Mullins was happy to hear that the crew saw him and his chiefs as a united team for that was exactly the type of image and reality that he was attempting to cultivate. By demonstrating his competence, by serving with his subordinates from the very outset and by acting on their behalf when required, he created a cohesive team that served HMCS *Montreal* well.

In addition to his department chiefs, CPO1 Mullins also paid considerable attention to his many subordinates. To the best of his ability, he ensured that their welfare was looked after insofar as time and resources would allow. First and foremost, he always made himself available. His door was always left open, thus indicating that all were free to enter his office and to speak to him on any matter should they so desire. Possessing a satellite phone in his office, he allowed his sailors to call home outside of working hours, rather than keep it entirely for his own use. Moreover, when making nightly rounds, he and the XO, who was a very personable individual, learned a great deal about the ship's company, troubles included, by taking advantage of each opportunity to speak with them. On many occasions, rounds that would normally take thirty minutes or so frequently developed into one or two hour conversations with the sailors. Such interest and concern paid off in the end, for as CPO1 Mullins himself acknowledged, "it helped" to create trust between superiors and subordinates and a feeling of belonging and community aboard ship.[10]

Protecting the welfare of his crew assumed many forms and was not restricted to simple acts of generosity or interest. At the start of the deployment, a member of the commodore's staff frequently attempted to have members of *Montreal*'s crew do certain things by stating, with absolutely no authority, that "the commodore wants it."[11] This particular individual used the power and prestige associated with his superior's rank in an attempt to influence the behaviour of others. One of CPO1 Mullins' chiefs, who was excitable and animated to begin with, but extremely competent and loyal, became so frustrated with this situation that he tendered his resignation in protest. Attempting to defuse the situation through levity and humour, Mullins asked the chief if he was going to swim home, for if he resigned, the Navy could not bring him back! On a more serious note, however, he confronted the individual who was constantly "slipping on the commodore's rank" and told him that if this manipulation did not cease, he would approach the commodore himself and explain the current situation.[12] CPO1 Mullins reasoned, owing to his own position and credibility, that his explanation would carry more weight, with the obvious implication being that his nefarious colleague would suffer the brunt of the commodore's dissatisfaction. In the end, the behaviour stopped and the impulsive chief decided to stay after all.

In sum, the time that CPO1 Mullins spent as coxswain of HMCS *Montreal* during Op Apollo presented a number of unique challenges that, at times, required innovative solutions. When faced with sudden and immediate change, he took the time to rationally explain the reasons behind such alterations and illustrated why they were absolutely necessary. More importantly, though, he attempted to improve morale whenever he noticed that it was lagging and to keep the crew focused on the tasks at hand by dispelling rumours and by encouraging his subordinates not to speculate themselves, lest they add further to the negativity. For him, ensuring the welfare of *Montreal*'s crew meant shielding them from unnecessary and unwarranted intrusions and, in general, by simply "being there" for those who needed advice or assistance. On many occasions, his wise counsel offered to both the CO and XO influenced their decision-making processes and ultimately contributed to making HMCS *Montreal*'s deployment on Op Apollo a success.

ENDNOTES

1 This chapter is based on an interview between CPO1 Ken Mullins and Craig Leslie Mantle, which was conducted in Halifax, Nova Scotia, on 21 July 2006.

2 During Roto 2, Commodore Dan Murphy served as the Task Force Commander in theatre; HMCS *Montreal* served alongside HMCS *Winnipeg*. See Richard H. Gimblett, *Operation Apollo: The Golden Age of the Canadian Navy in the War Against Terrorism* (Ottawa: Magic Light, 2004), 80.

3 During its deployment, for instance, crewmembers from HMCS *Montreal* boarded a freighter bound for Iraq and seized five 24-metre patrol boats that could have been used for military purposes. See Ibid., 78. See also, "Canadians Intercept Suspect Freighter," *Ottawa Citizen*, 1 November 2002.

4 Interview quotation.

5 HMCS *Iroquois*, HMCS *Fredericton* and HMCS *Regina*, under the overall command of Commodore Roger Girouard, deployed as Roto 3 between, at the extremities, February and September 2003. See Gimblett, *Operation Apollo*, 80.

6 For an account of this incident, see Major Larry McCurdy, "Operational Readiness – Operation Apollo – A Helicopter Air Detachment Commander's Perspective," in Colonel Bernd Horn, ed., *In Harm's Way – On the Front Lines of Leadership: Sub-Unit Command on Operations* (Kingston: CDA Press, 2006), 167-82.

7 The chiefs who served under CPO1 Mullins included the Combat Systems Engineering Chief, the Marine Systems Engineering Chief, the Combat Chief, the Supply Chief, the Air Chief and the Chief Boswain's Mate.

8 Interview quotation.

9 Ibid.

10 Ibid.

11 Ibid.

12 Ibid.

CHAPTER 6

THE CANDYMAN COMETH: THE LEADERSHIP EXPERIENCES OF CPO1 (RET'D) PAT EVANS ON OPERATION APOLLO, ROTATION 3

Craig Leslie Mantle[1]

As part of Canada's contribution to the international campaign against terror, Chief Petty Officer First Class (CPO1) Pat Evans deployed on Her Majesty's Canadian Ship (HMCS) *Iroquois* on Rotation (Roto) 3 of Operation (Op) Apollo in February 2003. After arriving back in Halifax from Roto 0, the *Iroquois* experienced a number of personnel changes in quick succession. Within a month of returning to Canada, a new Commanding Officer (CO) and Executive Officer (XO) joined the ship, while a good proportion of the crew were posted elsewhere, all of whom were duly replaced. CPO1 Evans, the new coxswain, joined at this time as well. With a relatively fresh complement now onboard, the Navy decided that the *Iroquois* would follow a "low-key" schedule for the foreseeable future in anticipation of a much-needed refit. To keep busy, though, the CO ordered four days of assisted work-ups followed by a port visit to Charlottetown, Prince Edward Island. Given the significant number of changes both above and below decks, the time spent at sea would begin the process of moulding the crew into a cohesive whole that functioned well under all conditions.

During this period, rumours began circulating with some frequency that the *Iroquois* would be tasked with yet another Op Apollo deployment. Faced with such a prospect, Evans soon realized that more than one hundred of *Iroquois*' sailors, who had recently returned from Roto 0, had not yet been posted-off of the ship, a procedure normally followed upon repatriation that provided time ashore, thus protecting personnel from being immediately sent back to sea. If the deployment did occur, the command team – the CO, the XO and the coxswain – would thus be faced with the situation in which a large percentage of the ship's company would be compelled to return to the Persian Gulf and the Gulf of Oman for a second time in the span of less than a year, with minimal time to decompress in between deployments.

The potential for low morale and frustration to sap the sailors' enthusiasm was both obvious and clear. In anticipation of sailing again, Evans began the difficult process of motivating his sailors, especially those who had just returned from overseas, so that they would be more eager and willing to deploy again should the need arise. To this end, he spoke to them at length about the importance of the mission itself and how their actions would contribute to international security. He also took pains to reinforce the fact that they had all enlisted in the Canadian Forces (CF) and thus were obliged to go where and when they were told. To take some of the edge off of these comments, Evans informed his sailors that the Navy, up until Op Apollo, had rarely been called upon to go into areas of conflict so they should all be grateful for such an opportunity.[2] The Navy was beginning a new chapter in its already-storied history and he encouraged his sailors to revel in the moment.

When it was finally decided that the ship would participate in Roto 3, the CO assembled all hands on deck and informed them of their imminent departure.[3] The next two weeks were spent completing the necessary pre-deployment clearances and ensuring that both personal and family matters were in order. Fortuitously, the command team had already begun many of these procedures when they first learned that the ship might be deployed. By anticipating such possible events, they were able to save valuable time, although this period was nevertheless hectic and busy for everyone onboard. For his part, CPO1 Evans spent much of this two-week interval coordinating the various outside agencies, such as the Military Family Resource Centre, the Judge Advocate-General and the Base Hospital, that needed to visit the ship to complete various clearance procedures. He also worked closely with his department chiefs by holding daily meetings and informing them of the type of work that would occur each day. Where possible, he told them of their upcoming schedule, thus ensuring that they could anticipate and prepare for their assignments. He relied heavily on his immediate subordinates during this time to ensure that their departments were in order and that their sailors spent as much time as possible with their families before leaving Canada.

With these initial difficulties more or less overcome, the *Iroquois* set sail from Halifax for the Adriatic. The transit, as usual, was to be spent completing work-ups that were designed to ensure that both the ship and its crew were operationally ready. Four days out, however, the *Iroquois* suffered a major accident when its helicopter crashed back onto the deck.[4]

Only two sailors suffered injuries, a fortunate circumstance as the results could have been much worse than they were. As a consequence, the *Iroquois* quickly returned to Halifax to be inspected and repaired if necessary. Once a survey of the flight deck revealed it to be structurally sound, the decision was made to proceed with the deployment as scheduled. While alongside, the damaged helicopter was removed and the aircrew departed as their presence was no longer required. With the mission continuing as planned, the only doubt that remained was when and where the *Iroquois* would receive another aircraft.

The psychological toll exacted by the accident, both immediately and in the aftermath, was extreme. Not only had a major incident occurred with the crash of the Sea King, but the ship's company was forced to say goodbye to their families yet again after being given leave while the fate of the *Iroquois* was being decided. In Evans' opinion, however, the crew dealt with the incident in an extremely professional and competent manner. Their earlier training had prepared them well to react to emergencies such as this. Recalling the incident years later, he observed, "The aft end of the ship and all mess decks were cleared in record time. Our section base teams were dressed and ready to do their job. The casualty clearing teams had the two injured personnel in hand and [they were] being looked after."[5] This accident, difficult as it was, at least proved to Evans and others that *Iroquois*' company was functioning as a team and that everyone was generally well prepared for their deployment. Rather than try to forget what had happened, Evans encouraged his sailors to talk about what they were doing at the time of the crash, what they thought was happening during the incident and how they had reacted. He especially wanted to know how the accident had affected them personally and what their families thought of the event. To ease some of the stress that many were naturally feeling at this time, he tried to reinforce with each member that their training had truly served them well. Despite the potential for a catastrophe, "... the crash served to further bring our ship's company closer together as a team. We realised that while things could go wrong, we were well trained and prepared to deal with them."[6] In a few days time, after the necessary repairs had been made, the *Iroquois* left Halifax again. As planned, work-ups occupied much of their time during the transit and, in Evans' estimation, the crew "really came together."[7]

Even before the ship reached its area of responsibility, CPO1 Evans was called upon to make a significant leadership decision that would ultimately set the ship's disciplinary tone for the entire deployment. After the work-ups had been completed, the crew was given overnight leave in Greece, the only stipulation being that they had to provide the name and telephone number of the place where they could be reached in the event of an emergency. Such precautions were deemed necessary given the anti-American (and thus, anti-coalition) sentiments of the local population. The following morning, however, four sailors were found to be absent from muster; they had, in effect, violated the very policy that had been designed to ensure their safety. CPO1 Evans reaffirmed his belief, after conferring with the CO and XO, that disciplinary standards had to be maintained and that they would be enforced in a fair and firm manner *for all* onboard. He consequently proceeded to investigate the four absentees (an officer, a petty officer (PO) and two junior non-commissioned members (NCMs)). After determining why they were late – many had simply decided not to return to the ship at the appointed hour so that they could further enjoy their time away – he recommended that everyone involved be charged with a service offence. In the end, all were found guilty and punished. The decision to proceed against the absentees, regardless of rank, set the disciplinary climate for the remainder of the deployment as the entire ship's company realized that standards had to be followed and would be enforced if they were not. More importantly, however, they also understood that no one was "above the law," that all would be treated the same on the point of discipline and that no favouritism would be shown.[8] Had Evans decided to charge only certain individuals, it is possible that the ship's company would have resented such a double standard and their performance may have suffered as a result.

Evans' leadership abilities were soon tested in theatre as well. In order to maintain an adequate level of force protection, the boatswains and naval communicators were tasked with the responsibility of operating *Iroquois*' .50 calibre machineguns. It soon became evident, however, that the number of such individuals was simply too small to adequately perform this task since their watches came around all too often. Consequently, with a plan already in mind to solve this dilemma, Evans consulted his chiefs to "find" an acceptable solution. During this meeting, one of his subordinates asked if his department could be made responsible for one of the guns. His sailors, so the chief offered, would not only ensure that the proper training was conducted, but would also create an acceptable duty schedule to ensure that this additional responsibility was divided equally amongst all. The remaining chiefs agreed that this was a logical and appropriate solution and soon volunteered their own departments in turn. Of course, this was exactly the plan that Evans had envisioned all along, however he was more than satisfied to see it come from his co-workers. The fact that they had volunteered for this duty revealed to him that the ship's company had achieved a high degree of cohesion. In the end, the four guns were divided amongst the various departments, with only one remaining with the original team of boatswains and communicators. Such an arrangement improved morale, despite the additional duties, as everyone aboard ship soon realized that all were sharing equally in this task and that no one individual or group was bearing the brunt alone. To be sure, it was not uncommon during the deployment to see a lieutenant-commander and an ordinary seaman sharing a watch in the middle of the night. By allowing the chief to suggest a novel solution, Evans furthered this individual's commitment by allowing him to assume a leadership role and by giving him ownership over his own idea, all of which proved to be extremely motivating.

Certainly, the leadership challenges did not end there. After arriving in their area of operations, the *Iroquois* was tasked with the responsibility of boarding foreign vessels in search of Al-Qaeda terrorists and their weapons.[9] To this end, members of the boarding party received intensive training, both during the transit from Halifax and while in theatre. On two separate occasions, however, two different members of the boarding party suffered an accidental discharge of their weapons. The subsequent investigations revealed that while the discharges were not wilful, they were entirely avoidable in that momentary lapses in concentration had caused the weapons to fire. Luckily, no one was injured on either occasion. Because of their training, equipment and duties, members of the boarding party had earned a reputation of being an "elite" of sorts amongst the other sailors. When the two discharges occurred, however, their sailing mates began to question their abilities and to wonder not only if the training that they were receiving was really of the proper type, but, more damaging, if they really were the most suitable candidates for the job. If they truly were elite, why, then, were these accidents occurring?

After determining the cause to be nothing more than carelessness, Evans informed his department chiefs that charges would be laid against the offending sailors. At his daily meetings, he made the reasons behind the discharges known to all in order to put an end to the speculation and rumour that was quickly developing onboard.[10] Once the chiefs were so informed, they in turn communicated the reasons to their own subordinates, with the results that much of the idle gossip stopped in due course. Communicating relevant and useful information truly helped Evans end what could have developed into a significant morale issue if it had continued for much longer. Rather than trying to hide the reasons behind the discharges, he made them freely known to all in the hopes that they would be corrected in the future.

Although certain events during the deployment caused considerable excitement and concern amongst the ship's company, such as the crash of the helicopter and the accidental discharges, life in theatre eventually became routine and somewhat monotonous. For members of the boarding party, though, a fair amount of stress and anxiety accompanied their duties as they faced the unknown every time that they left the relative safety of the *Iroquois.* To reduce the negative influences of boredom and exhaustion, CPO1 Evans and the rest of the command team tried to do as much as they could for the ship's company. On many occasions, they held Banyans, a relaxing break from routine in which the sailors pursued a variety of activities. As before, Evans consulted his chiefs and, through their collective efforts, they hosted a number of events that were particularly noteworthy. During one gathering, the sailors had the opportunity to listen to a band, play basketball, sun themselves, go skeet shooting, enjoy a barbeque and hit golf balls into the sea … all in one day! The crew of the *Iroquois* also participated in "No Days," usually on a Sunday, in which as few duties as possible were performed, excepting those of absolute necessity or those occasioned by operational commitments. When alongside in foreign ports of call, the command team tried to give everyone as much liberty and latitude as possible, such as overnight leave.

CPO1 Evans also faced a number of ethical dilemmas throughout his deployment. During Roto 0, the Vice-Chief of the Defence Staff established a policy concerning the consumption of alcohol while in theatre that, as events would prove, became exceedingly unpopular with the sailors. In sum, no CF personnel were to consume more than two beers, two ounces of alcohol or two glasses of wine in a twenty-four hour period. Such restrictions held whether the member was onboard, in a foreign port or given overnight leave ashore. Although the region was Muslim, all major hotels in Dubai and the Seaman's Club in Jebel Ali served alcohol. The problem that Evans faced was one of enforcement. If a member appeared drunk, it would be easy to assume that he or she had had more than two drinks. Since they had apparently violated the policy on alcohol, disciplinary charges could then be laid. In order to prove someone had exceeded the limit, however, a witness would be required to state that he or she saw the individual in question consume X number of drinks. In Evans' opinion, this sort of behaviour would seriously undermine the trust and cohesiveness of the crew. As a consequence, he chose not to "lurk in the bushes" and count the number of beverages being consumed by any one individual. Instead, he tried to keep a high profile at the Seaman's Club, often engaging the groups of crewmembers in conversation and, upon his departure, saying something like, "Gentlemen, you know the policy…be safe." Evans believed that everyone realised that he would not hesitate to lay charges against them should an incident involving alcohol occur, especially given his earlier indications to this effect while the *Iroquois* was in port in Greece. Certainly, he was not naïve enough to think that no one ever drank more that two drinks in one day, however he chose to passively enforce the policy rather than to do so aggressively. Fortunately, during their time in theatre, no incidents involving alcohol occurred.

A welcome change to this policy was received about a month prior to the departure of the *Iroquois* for Canada. The two-drink maximum was still in effect for those who went ashore, however while onboard, sailors could consume more than two drinks under the normal mess rules. To ensure that no incidents occurred in which members consumed several drinks onboard and then went ashore, Evans instituted a unique policy at each of the messes. An ammunition box was kept at each bar. It had a lock on it with a slot large enough to accept a sailor's identification (ID) card. If a member wanted a drink onboard, he first had to surrender his ID card to the bartender. As the bow staff checked ID cards before allowing personnel to go ashore, this policy ensured that all members who had been drinking were safely onboard for the night and could not leave. The Duty Coxswain opened the box the following morning and distributed the ID cards to their respective owners. This arrangement worked well in that it enforced the alcohol policy while at the same time keeping the sailors safe and allowing them some of their freedoms.

In order to keep a watchful eye on the morale and welfare of the crew, in effect to learn what they were thinking and feeling, Evans used a variety of techniques to learn if the ship was "happy" or not. When departing each port, for instance, he habitually visited everyone onboard and distributed candy, certainly not an easy or inexpensive feat, as more than 300 sailors served onboard the *Iroquois*. While a nice gesture, it also provided an excuse to approach the sailors and to chat lightly and informally about "things." Because of his approach, the crew nicknamed him "The Candyman," a name he thought suitable and appropriate. He also maintained a strict open-door policy where all were invited to speak to him about their difficulties. Making himself available reassured the crew that someone would listen to and act on their problems, whatever they may have been. Treating everyone in a pleasant manner and with respect – first names were not tolerated for Evans believed that such informality would weaken discipline – created a positive, family atmosphere aboard ship. He insisted, for instance, that everyone, officers included, greet one another when passing in the flats with a "Morning Sir," a "Good afternoon Petty Officer," or a "Good day Leading Seaman."

Such efforts proved quite beneficial for he eventually learned of a number of difficulties that were developing amongst the small contingent of female sailors, which numbered 34 or so out of a total complement of around 330. In particular, one NCM was having some difficulty with her job. In her mess, some of the sailors were siding with her, saying that "the system" was too harsh in making her meet certain performance standards, whereas others thought that "the system" was too lax for not removing her for incompetence. Her position required that she know the ship "inside and out," or in other words, to understand its many functions and capabilities. Sailors in her position were normally given three months or so to learn their responsibilities and then to sit on an examining board where they would be qualified to carry out their duties without supervision. This individual, though, had spent more than four-and-a-half months in her position and had not yet proved her competence; she had not even attempted the exam as she lacked confidence in her knowledge. Because she was not qualified, she disrupted the ship's rotation schedule as other individuals were forced to supervise her while she was on duty. Had she proven herself competent, she would have been permitted to act alone, thus reducing the extra demands on her crewmembers. The fact that she was in a position of responsibility and leadership, but had failed to ensure that she was fully capable, caused a good deal of resentment amongst some and this bitterness eventually began to negatively affect morale. In an attempt to come to an acceptable solution over the course of many weeks, this particular sailor was made to stand double-bank watches with more senior personnel in an effort to "learn the ropes." It was hoped that because these watches were more frequent than those stood by her peers, she would be motivated enough to sit the examining board. A significant effort was truly made to raise her confidence. After these efforts failed to have the desired effect, Evans, in his capacity as coxswain, approached the CO, informed him of the present difficulty (the captain was generally aware of the situation already) and offered his advice on how to proceed. Eventually, the command team decided that it would be best for all concerned to return this sailor to Halifax. In due course, she was sent back to Canada and the remainder committed themselves to "picking up the slack" for there was little else that could be done. Morale, not surprisingly, improved once this individual had been repatriated.

Another situation soon developed that forced CPO1 Evans to rely heavily on one of his subordinates to reach an adequate solution. Several female sailors had earlier confronted him with their concerns about the conduct of one of their co-workers. In particular, they mentioned that they believed this individual was leading an "alternative lifestyle." While certainly not a problem for him, or officially for the military, he was forced to investigate the matter. Her alleged homosexuality was not the impetus behind his enquiries, but rather the fact that her behaviour toward the other sailors might have constituted an illegal act, namely sexual assault. Evans was thus duty bound to investigate. After conducting a background check on the sailor in question, he asked a female warrant officer (WO) with whom he was serving to interview those who were making the complaints and to report her findings to him, whereupon he would decide on how to proceed.

Once the interviews had been conducted, it became readily apparent that the sailor was not homosexual, but merely "close," that is, she frequently stood within people's personal space and was very comfortable with her own sexuality, so much so that she frequently appeared naked or semi-naked in regions of the ship where other female sailors usually covered themselves, such as in the wash areas. Because quarters were tight and because she often wore very little, if anything at all, she frequently brushed-up against her companions and they disliked this close contact intensely. To inform the individual in question that her "lifestyle" was causing considerable angst and discomfort amongst the other female sailors, Evans arranged for the WO to mediate a meeting (which he did not attend) so that all could voice their concerns. The sailor, when confronted by her peers, appeared somewhat shocked and embarrassed by the entire situation, but listened attentively to everyone's opinion. In the end, she changed her behaviour and no further complaints were raised.

In this particular instance, rather than attempting to deal with the situation on his own, which given the nature of the matter would have been uncomfortable for all involved, Evans enlisted the help of a female subordinate. He reasoned, correctly as events would prove, that the women aboard ship would relate better and more freely to another female, rather than with an "old salt." With this being said, however, he always made sure that everyone understood that the WO would brief him eventually as to what was said during each meeting. By so doing, he remained fluent with the situation, but saved both his sailors and himself considerable embarrassment and frustration.

While successfully resolving many of the challenges with which he was faced, there were a few that could not be settled despite his best efforts. The *Iroquois* originally did not plan on sailing with a helicopter detachment, although immediately before deploying, it was decided that they would take one along for extra flexibility. Consequently, the aircrew from HMCS *Fredericton* was removed from their present berth and posted to the *Iroquois.* Their resentment at the sudden change was more than evident. Soon after the crash, as if this was not quite enough, they were removed altogether while the ship's fate was being decided. In order to provide the Iroquois with an air detachment, another helicopter was later transported directly to theatre and the same aircrew, for the most part, was flown overseas to meet the ship. In an attempt to ease their various transitions and to calm their frustrations that naturally resulted from being posted-in, posted-out and posted-in again in theatre, Evans tried to give them better accommodations than what they were normally "entitled to." Petty Officers Second Class, for instance, were given bunks normally reserved for Petty Officers First Class. In addition, he tried to team build by linking members of the air department with certain *Iroquois* crewmembers who shared similar interests. In this instance, learning about his subordinates proved doubly valuable. He made a concerted effort to introduce those individuals, for example, who played guitars and who had an interest in music. Finding colleagues with similar tastes, so he reasoned, would allow friendships to be formed and would provide a diversion from the uncertainties of daily life.

Despite these attempts at team building, the air department rarely volunteered for any extra duties during the remainder of the deployment. The other departments, in contrast, freely gave of their time to help work the .50 calibre machineguns, to manually load the ship with provisions while in port and to assist the ship in coming alongside. When extra personnel were required from the air department, CPO1 Evans was oftentimes forced to order them to assist, whereas the sailors tended to share the burden willingly. He later attributed this reluctance to differences in culture, at least partially. In his opinion, in the Navy, the entire ship's complement, officers and sailors alike, journey into harm's way together. All share the same risks, so there is naturally a greater willingness for everyone to share the extra labour and duties when required. In the Air Force, however, NCMs are largely responsible for ensuring the airworthiness of the platforms that officers operate. All are not exposed to the same degree of danger, so there exists a greater reluctance to perform those duties that are judged to be the preserve of others. When attempting to find additional personnel to operate the .50 calibre machineguns, for instance, the air department sent four officers only, as it was perceived that such activities fell into their domain alone.[11] With this being said, however, Evans had worked with other air departments throughout his career that freely gave of their effort and who could really be considered part of the team. Cultural differences aside, their unwillingness to assist may also have resulted from their treatment during the last few months: they had been removed from the *Fredericton*, removed from the *Iroquois* after the crash and then transported overseas to join the *Iroquois* again in the middle of the deployment. In essence, the air department had not been with the crew during work-ups and so some of their reluctance to volunteer may have resulted from the fact that they had not served with the ship from the outset. Being absent during these critical trials, the bonds of cohesion and comradeship that normally encourage personnel to assist others when in need lacked a certain amount of strength.

All in all, CPO1 Evans experienced an eventful deployment in which leadership challenges and ethical dilemmas forced him to act on his own initiative and to rely on the assistance of others. His efforts to monitor and ensure the welfare of all of his sailors took many forms and provided him with a forum to more fully comprehend their attitudes. When faced with non-life threatening decisions, Evans frequently allowed his subordinates to offer their suggestions and to advise on the way ahead, thus further reinforcing their commitment to the larger team environment. Although many of these techniques had been learned and refined during years of continuous service, cultural barriers proved insurmountable and resistant to traditionally useful approaches. Importantly, the position of privilege and access traditionally held by the coxswain allowed him to advise his superiors on the next course of action in certain circumstances. Like other sailors aboard HMCS *Iroquois*, Evans experienced a successful rotation and overcame significant challenges owing to his professionalism and dedication.

ENDNOTES

1 This chapter is based on an interview between CPO1 (ret'd) Pat Evans and Craig Leslie Mantle, which was conducted in Halifax, Nova Scotia, on 25 July 2006.

2 In Evans' opinion, the greatest difficulty for most of these sailors was not deploying again on a second rotation, difficult as it was, but rather convincing their spouses that it was necessary that they return. The willingness of their family to endure yet another six or seven month separation contributed materially to *Iroquois*' eventual success on its subsequent deployment.

3 Being told as soon as possible what their mission would be once in theatre also helped to motivate the sailors.

4 For an account of this incident, see Major Larry McCurdy, "Operational Readiness – Operation Apollo – A Helicopter Air Detachment Commander's Perspective," in Colonel Bernd Horn, ed., *In Harm's Way – On the Front Lines of Leadership: Sub-Unit Command on Operations* (Kingston: CDA Press, 2006), 167-182.

5 CPO1 (ret'd) Evans to C.L. Mantle, 8 Sep 2006, via e-mail.

6 Ibid.

7 Interview quotation.

8 Ibid.

9 For further information concerning naval boarding operations, see Richard H. Gimblett, *Operation Apollo: The Golden Age of the Canadian Navy in the War Against Terrorism* (Ottawa: Magic Light, 2004), 92-93.

10 As an aside, one of the sailors involved in these incidents had originally lied about how his discharge had occurred. He stated that the weapon's cocking handle had caught on a wing nut on the top of an ammunition locker, thus causing the weapon to discharge due to a "slamfire," something that is possible but very unlikely to occur. The XO was an expert in small arms and launched an immediate investigation due to the importance of determining if the *Iroquois* had a bad lot of ammunition. The results of his investigation were potentially far-reaching, for if the ammunition was found to be defective, it would have to be quarantined throughout the CF to ensure that no further misfires occurred. After several days of testing both the ammunition and the weapon, the "slamfire" could not be duplicated. Evans had re-interviewed the sailor in question several times in the days following the accident in order to learn as much as possible about how the discharge had occurred. Each interview was done under caution so that the member had the right to refuse to talk if that was what he wanted. The sailor chose not to invoke his right and it soon became evident that something was truly bothering him, but that he just could not say. Evans' philosophy on service related offences, and he told the sailor this, has always been, "If you didn't do it, stick to your guns, however if you did, step up, take your punishment and get on with the job of being a serviceman." The next day, the sailor approached Evans and relayed the true sequence of events. In reality, his weapon had been accidentally loaded when he dragged it across the ammunition locker, however he had panicked when he cleared the round and had not followed the proper procedure. The member in question was a good sailor who had made a mistake. He admitted that the worst part of all of this was that he had lied to both the XO and Evans in the beginning. He was duly tried and convicted, but remained with the ship for the full deployment. Evans to Mantle, 8 Sep 2006, via e-mail.

11 The foregoing represents the opinion of CPO1 (ret'd) Pat Evans only and does not necessarily reflect the perspective of the present author.

CHAPTER 7

LESSONS LEARNED FROM A CAMP SERGEANT-MAJOR OPERATION ATHENA, ROTATION ZERO, KABUL, AFGHANISTAN

T.J.P. Mugford

The mission on the battle group to which I was attached was to provide security and stabilization in support of the Afghan Transitional Authority (ATA) as part of Operation Athena, Rotation (Roto) zero. My specific responsibilities included the defence and security of the camp interior. Additionally, I was responsible for quality assurance with regard to the construction and set up of the camp. This task included coordination, liaison and management of camp workers who, notably, derived from a variety of nationalities including locally engaged employees, Indians and Nepalese. Moreover, I was in charge of coordinating all visits from very important persons (VIPs) and very, very important persons (VVIPs) in terms of security requirements, housing and parades. As well, as required, I ensured the enforcement of camp policies, orders and directives.

Situated in Kabul, the environment was extremely hostile. Firstly, there was marked instability at the local and national government levels. Secondly, there was continuous Al-Qaeda terrorist activity, as well as attacks and threats on a regular basis sanctioned by drug lords. Moreover, there were an estimated sixteen to twenty million mines in and around the countryside. As a result, clearly mines remained a constant threat.

Given the large scope of the mission – to provide security and stabilization to the ATA – and the hostile environment in which we were forced to operate, it is not surprising that many leadership challenges occurred. Foremost was the question of how to influence workers from a variety of different ethnic and cultural backgrounds to perform tasks that conformed to a military based plan and execution style, for example, camp construction. It was imperative to first understand and accept the cultural differences among the civilian workers in order to be able to forge a way ahead. Our goal was to construct the camp in the most expedient manner possible. In order for us to accomplish this task we needed to treat the local workers in the way that would best further our objectives. To facilitate this end, each cultural group was treated uniquely in an effort to encourage their full participation in the project. For example, the Indians and the Nepalese did not share the same work ethic. This disparity created a number of conflicts. Subsequently, I separated their living and work areas and, in so doing, avoided any further incidents. Moreover, I quickly made contact with and befriended the leaders of each cultural group and involved them in the decision process when dealing with any and all issues that arose. This helped them feel at least partially in control of their environment. For example, the local engaged employees and the Indians had different hygiene habits compared to the rest of the camp personnel. The issue became so bad that the morale of many members of the camp was affected. Several complaints were forwarded to my attention and recommended courses of action were put forth. In the end, the groups in question were given a segregated area where they could carry out their hygiene habits in accordance with their cultural ways and customs.

Another issue that arose was how best to deal with the day-to-day concerns and requests from military units within the camp and, in particular, their attitudes towards the Camp Commander's priorities versus their own quality of life issues. Each, according to the different cultural groups, had their own precedence. For example, the units, for the most part, felt that permanent accommodations and quality of life issues were of paramount importance. Conversely, the Camp Commander prioritized force protection and camp defence and security issues. On several occasions, unit representatives would refer back to Bosnia and the low quality of life that existed in the military camps there as a means of substantiating their requests. Subsequently, and unfortunately, during the initial part of the tour, the mission itself was misunderstood by many. To solve the problem, I elected to send electronic and hard copy reminders to the group on a regular basis and at every weekly Camp Sergeant-Major Conference to re-emphasize the Commander's priorities. I also frequently inspected the areas of the camp that still required defence construction and prioritized that work with the contractors. The message was, unfortunately, only finally driven home with the first rocket attack on the camp and from that point forward, security and force protection were no longer issues or open for discussion. A serious lesson was finally learned and accepted by one and all!

A third challenge that occurred was how to deal with a lack of clear direction in the absence of a transparent chain of command. Unfortunately, this was our situation for the first month in theatre. In particular, the issue became one of how to deal with the tense and stressful work environment that this lack of clear leadership created. For instance, a lieutenant-colonel was filling the job of a major and the major was subsequently filling the job of a captain and so on. Thus, it was very hard to define the actual chain of command, particularly because everyone was in search of a "raison d'être" and trying to fulfill the wishes of the Commanding Officer without clear Terms of References or clear direction. As a result, some individuals took liberties with regards to tasking personnel or deciding priorities of work for the camp. To help solve the problem and in order to perform my job as effectively as possible, I immediately requested an interview with the Commanding Officer and proceeded to express my concerns and asked for clear and defined terms of reference including a distinct chain of command. The result was favorable, but personalities always seem to get in the way and one must make compromises in order to ensure the mission is not jeopardized and always first and foremost on everyone's mind. Notably, a last minute decision was made to create a National Support Element (NSE) that included an additional unit and command structure for the mission and, in particular, the camp. Arguably, this addition was not necessary and created redundancy with regard to support to the fighting troops and, more importantly, caused delays and confusion with providing that critical and essential support.

A fourth concern was how to deal with the frequent demands for the housing of media groups and personnel from allied forces within a camp that was already overfull and behind schedule. Specifically, it was a challenge to provide them with secure areas for training, rehearsals, work and accommodations. To solve this problem, I simply applied the estimate format (i.e. aim, factors, courses open and plan) to the decision-making process. I also requested realistic compromises from the permanent members of the camp and the allied force or media group in question. The final outcome in each case was total success. Importantly, this challenge also involved understanding and respecting cultural and ethnic differences. These differences were always considered in the equation.

At the unit level, we were also faced with a number of challenges. The following were the most notable:

- dealing with a multicultural workforce and the issues that arose from it;

- dealing with the constant threat of mines and rocket attacks in and around the camp;

- dealing with the uncertainty of the stability of the local government and the impact it had on the sense of security of those members who frequently dealt with the local populous; and

- dealing with the weather, in particular, the desert like conditions and harsh winter and the impact the environment had on essential vehicles and equipment, as well as troop morale.

In conclusion, the unit performed extremely well, especially considering it was the first rotation and a first mission of that intensity for most of the soldiers and leaders. I firmly believe that units should not deploy on a Roto 0 with any expectations that quality of life issues will take precedence, or even be truly considered, until the mission itself is up and running in accordance with the Commander's intent. In addition, this should be made clear during pre-deployment training and by the highest to the lowest levels of leadership. The mission has to always remain the primary focus and this must be instilled in the minds of every member of the operation. Nonetheless, quality of life issues, if feasible, should be considered, especially as the mission progresses, and implemented as required. At no point, however, should they have an impact on the mission tempo itself.

CHAPTER 8

LEADERSHIP CHALLENGES DURING FORMED AND COMPOSITE UNIT DEPLOYMENTS

Thomas J. Secretan

As an Air Force Chief Warrant Officer (CWO) I have been fortunate enough to be the CWO for two overseas deployments: Operation (Op) Kinetic in Kosovo, June to December 1999 and Operation Athena, Camp Mirage, July 2003 to January 2004.

The Canadian Air Force traditionally deploys its forces in two different ways. The first is as a formed unit. This is typically how the military deploys on operations. Basically, a ship/regiment/squadron is chosen to deploy on an operation and does so as such, with its command structure intact. The second manner that the military deploys its forces is as a composite unit. In these instances, personnel from several ships/regiments/squadrons are brought together to form a new unit, sometimes in the theatre of operation. The unit's command element is then made up of different members from the different units.

For Operation Kinetic, my official title was Kosovo Rotary Wing Aviation Unit (KRWAU pronounced Crow) CWO. The vast majority of the approximately 176 personnel came from 408 Tactical Helicopter Squadron, of which I was the Squadron CWO. The unit deployed as a formed unit. On the other hand, for Operation Athena, my position was the Theatre Support Element (TSE) CWO. My home unit was 402 Squadron and I became the TSE CWO shortly after my arrival at Camp Mirage. The TSE was made up of over 250 personnel from 13 Air Force wings, and was also responsible for a cadre of Navy and Army personnel who provided direct support to the ships in the Persian Gulf and Task Force Kabul (TFK) in Afghanistan. Both deployments were Rotations (Roto) Zero.

The leadership challenges of deployments are complex. As a leader, I had to be able to identify the leadership challenge and adapt or modify my leadership to meet those situations. It is a comparison of those different leadership challenges and methods that I now want to go into. Before I get started, however, I must state that this is not a comparison of the different ways the Air Force deployed its personnel but rather a discussion on the different leadership challenges I faced while in theatre. Both types of deployments have positive and negative aspects and either can be effective depending on the deployment.

FORMED UNIT

The Mission

For my tour in Kosovo, 408 Squadron was tasked by 1 Canadian Air Division through 1 Wing in Kingston, to provide eight CH-146 Griffon helicopters that would be under operational control (OPCON) of the 4th Armoured Brigade of the British Army. Our over arching mission was to support the ending of the ethnic cleansing in the province of Kosovo in Serbia. Our tactical missions were multiple and varied, including troop insertions, observation, over watch, command and liaison, and medical evacuation on a twenty-four hour a day, seven day a week basis.

Communication

On receipt of the formal tasking for Operation Kinetic, the Commanding Officer (CO) formed up the entire Squadron and briefed them on the tasking in general terms. Immediately after, a senior staff meeting was held, orders were issued and some questions were answered. Due to the short reaction time, the CO then reinforced the leadership challenges that we were faced with. He reminded everyone of their position in the chain of command and finished by stating emphatically that the only way this operation was going to work was on the backs of the Senior Non Commissioned Officers (Sr. NCOs). They were the "how" people and the flight commanders needed to keep to the "what and when."

Upon returning to my office I was met with my first big leadership challenge. Three of my flight Warrant Officers (WOs) were waiting for me. They all had the same concern. Our Squadron had already been tasked to be the first helicopter unit into Bosnia, Operation Palladium, and, in fact, some of our squadron were currently on that deployment. Although our actual deployment to Kosovo would occur after the return of the Bosnia rotation, it meant that our resources to generate another tour, especially in personnel manning, was not up to strength. It was obvious the WOs were relaying the concerns of the Non-Commissioned Members (NCMs).

Consequently, I asked them to gather up the NCMs in the squadron briefing room. I would give them a chance to express their concerns. At the end, I explained to the airmen and airwomen why our unit was chosen, and although the task was enormous, we were the ones to get it done and done right. I reinforced the CO's comments in regards to the Sr. NCOs. I reminded them that this type of operation was exactly why they joined the Canadian Forces (CF) in the first place. I finished by stating my utmost confidence in their abilities. I then instituted a morning parade for all the NCMs during which a flight WO or myself would inform them of any and all changes in the plan. At that time, they would be allowed to direct questions to me. The morning parade initially met with some resistance from some of the flight commanders who thought it was a waste of time and an infringement upon their chain of command. The CO backed me on this issue and, in the end, the parade soon turned into a mission update briefing for all squadron members.

Once deployed to Macedonia, the ability to fully brief the squadron members became impossible. In this case, the CO would call in his flight commanders to explain the situation and to ask for options. He then made his decision and provided a rough outline for implementation. The plan was subsequently actioned with the details worked out on the fly. This meant that everyone had to have absolute confidence in his or her superior's abilities. Detailed planning was a luxury we no longer could afford. Consequently, explanations to the airmen and airwomen were not forthcoming. Action over explanation was the rule. This hierarchy caused its own set of personnel problems. Clarity of orders became essential.

Later in the tour, the unit had access to telephones and the Defence Information Network (DIN). These technologies brought a wealth of information but also an inevitable over attachment to the computer. The habit of sending directions via e-mail was quickly creeping back into our day-to-day business. To ensure the flight Sr. NCOs and WOs did not become too compartmentalized, I held a morning coffee break for all the KRWAU Sr. NCO and WOs. These were mandatory, except for emergencies.

Team Building

The majority of the KRWAU were from 408 Squadron with one or two augmentees from other tactical helicopter squadrons. The Airfield Security Flight (ASF) was the exception. It was composed of 43 military police officers from across the CF.

The CO and I insisted that these members be brought to our location to complete their pre-deployment training as an internal flight of the squadron. They where our first line of defence and it was essential that they and the rest of the unit train and work as a team. This team building was particularly hard for the ASF. Not only did they have to fit in with 408 Squadron, they also had to build their own internal ASF team.

Nothing builds teams like shared hardships and, in our case, the pre-deployment field exercise was the right tool at the right time. The exercise was mentally and physically very demanding. Eighteen hour days were the norm for everyone and Camp Wainwright in the spring provided a complete range of weather from hot sunny days to rain and snow and everything in between. The beer call (still an invaluable tool for team building) after the post exercise wash-up demonstrated all desired team building results had been achieved. The unit personnel were mingling freely and not remaining in their own flights. The war stories abounded with accompanying laughter. A deep sense of accomplishment and, more importantly, high unit pride was in the air. The KRWAU was good to go!

Ethical Dilemmas

Like most teams, the KRWAU was made up of a group of individuals who were working towards a common goal. Almost to a person everyone understood the importance of our mission. One individual, however, who was having some grave concerns, approached me. He felt that the situation in Kosovo was an internal problem to Serbia and that international intervention was wrong. We discussed this at length and he explained his point of view. Many of his points had merit and he had obviously been thinking about this issue for some time. This member was senior in rank to me and in a key unit position.

I had to weigh the member's misgivings and his personal ethical dilemma against unit cohesion, safety and operational effectiveness. We were too close to departure to train a replacement without a huge impact or delay to the mission deployment. My objective was simple: get the member to agree with and support our mission. If I did not succeed, then I would have to go to the CO and press for a replacement. He was in a key position and if a leader is not on board and committed to the mission, then it becomes infectious to the rest of the flight and the unit. I knew from our earlier discussions that trying to persuade the member through discussions of such things as international laws and United Nations (UN) mandates was not going to be effective. Rather, my tactic was to appeal to his morality in an effort to find a solution to his ethical dilemma. My final argument was that, as Canadians, we believe that political and/or religious differences are not a reason to kill. In fact, it was the obligation of the strong to protect the weak. As a country and a people we could not, and would not, allow another Rwanda to happen. We agreed that because of his being an airman and a Christian, the issue of genocide over-rode his interpretation of international law. This approach provided the desired results, and the member became an enthusiastic and determined leader committed to the mission.

Mission Change

There is an old saying. "No battle plan survives first contact with the enemy." Our battle plan was no exception. We were originally scheduled to pre-deploy to Prelip, Macedonia. It was expected we would have about three weeks to sort ourselves out and prepare for our movement forward into Kosovo. The problem was that no one had informed the Serbians of this plan. With in a few days of our arrival, it became obvious that our jump off time would be moved up. The squadron would be split in half: the Flying and Headquarters (HQ) flights moved forward to Skopje Airport, about 30 kilometres from the border with Kosovo; and the Support and Maintenance flights remained in Prelip. The squadron was neither manned nor equipped for a split operation. Coupled with that problem, some of our sea containers had been repacked prior to shipment and now our checklists did not match the sea container's serial numbers. Finding an aircraft part was impossible, so all 28 sea containers had to be emptied and repacked. The amount of work that lay before the unit was staggering!

The flying time between the different sections of the unit was only about 30 minutes but ground travel was a minimum of 2 1/2 hours, half of it on a narrow two-way mountain pass that provided every opportunity for disaster. The officer commanding (OC) the Support Flight became the officer commanding the rear area while the Commanding Officer (CO) moved forward with the HQ and Flying Flights.

The second major bump in the battle plan was our proposed location in Kosovo for when the air war ended. The KRWAU was to seize ground next to a taxiway at Pristina Airport, yet no one had informed our new "partners in peace," the Russian army. They had moved through Serbia after the end of the bombing campaign but before the coalition forces were allowed to enter Kosovo in accordance with the ceasefire agreement. This provided the Russians with the opportunity to secure a large portion of the Pristina Airport for their base camp. With the number of coalition forces moving into Kosovo, real estate quickly became limited. Consequently, we had to fit in reconnaissance (recce) flights between other missions, thus greatly increasing the workload.

A few days after flying operations in Kosovo had begun, a location for the Squadron was found. It was an abandoned crop dusting airport with some war-damaged buildings. Orders were issued and the KRWAU prepared to move into country. Although hostilities had ceased, intelligence indicated that small factions were not happy with the North Atlantic Treaty Organization (NATO) intervention and that all precautions for convoy safety were to be taken for the road move. The flying flight would join the main body after the basics of the new camp were set up. All KRWAU personnel who were moving on the ground were fully briefed. I would lead the advance party in with members of the ASF. The main body would join us the following day, with the aircraft arriving the day after that.

These two major changes in our plan were not the last. In fact, very little of our original plan remained intact. Everyday was met with continual changes. The key to the operation was flexibility, not only in our mission, but also in our own personal ability to adapt. The most important skill in a fluid environment is the ability to recognize that a change of plan is frequently needed in order to meet mission requirements. The second skill is to have everyone believe that a change of plan is part of the plan. It is essential that the unit members believe that someone knows what is going on. There are times when this may not in fact be the case, however, on these occasions, the CO and CWO must rely on their acting skills.

They must never show doubts, worries, concerns, or misgivings about the mission; only a calm, composed, determination can be projected.

Morale

After our initial deployment to Macedonia, following the squadron division between forward and rear echelons, I became aware of an underlying morale problem. The forward personnel felt they where roughing it in tents on the infield at the Skopje Airport, while the support personnel had an easy go in hardened barracks in Prilep. Conversely, the support element personnel left behind felt abandoned and out of the fight. Moreover, they did not consider that they were living in the lap of luxury. The aircrew operations camp was about 150 metres from the active runway in Skopje and with the amount of air traffic, uninterrupted sleep was impossible. Moreover, the temperature was forty degrees Celsius above freezing; the food was provided by the British; and the showers were across the airfield and unserviceable. Unfortunately, however, a growing "us versus them" syndrome was developing between the two echelons.

I felt a need to improve the morale of each group. In order to do so, I would need to improve their living situations. One of our back up plans in Canada was to purchase PVC piping, showerheads, water pumps and a small water tank so that, if required, we could build our own shower facility with some modular tentage. I made it a priority to find this equipment from the mislabelled sea containers and had it moved forward and set up as soon as possible. Additionally, on my daily runs from the support camp where I was working, I would stop and buy cold soft drinks and chocolate bars for the forward flight members. Later, when I found a store closer to the airport, I purchased popsicles.

During my time in the support camp, I explained to the Sr. NCOs and flight WOs that there was just no room for the whole squadron to be deployed forward and that support would have to continue with the long distance commute. I requested the CO make a trip back to the support camp to re-affirm that the rear echelon was still part of the team and their contribution was essential. This may seem like "molly-coddling" the troops and even a waste of the CO's time, but morale is the heart of a fighting unit and ours was already slipping and we had not even got into Kosovo yet. Furthermore, we still had at least five and a half months left of the tour.

One of my last morale problems came close to the end of the tour. The unit was to have a parade and receive their NATO Kosovo tour medals. The medals were to be presented by the Chief of the Air Staff (CAS). The problem arose when it became known that the CAS would fly in, present the medals and depart for the CF-18 Unit Task Force (TF) in Aviano, Italy after spending only three hours on the ground with us. Some of the members of the unit had determined that in six months, the CAS had visited TF Aviano on more than one occasion and would be remaining overnight (RON) with them again. This rumour quickly spread through the unit and many of the unit members felt the CAS should have spent more time on the ground with one of his more forward combat units. Even the new Governor General had remained overnight with us.

I was quite surprised how upset most of the KRWAU members were over this issue considering the repatriation of the unit would commence in five days. Unfortunately, I had no way to fix it and the best I could do was to keep the members focused on the important issues at hand. When I had the chance, I explained that the CAS's time is not his own and that it was not our job to question his schedule. Regardless, that night in camp was one of the lowest for morale I had seen during the whole tour.

Fear

Generally everyone was eager to move forward into Kosovo when it became time to advance. Nonetheless, a few members were very concerned about the pending move. I spent the evening prior to our advance walking the camp and chatting with unit members. It was essential that this move forward be treated as a normal administrative move, as if we were on exercise in Canada. In all actions, I tried to impart a sense of normality.

A couple of the members approached me directly with their concerns for the coming days. They knew I had already been in country on recce flights and had a good understanding of what was coming. We went to the mess tent and we chatted over coffee. I eased the discussion into the pre-deployment training, particularly the practice road moves we had done and the immediate action response drills. I reminded them how well we did at that and assured them that like any team, training and practice was the key. I kept the conversation on the light side and injected humour when I could. Like any old CWO, I told them of my own experience with fear. In the early 1980s, I had been a young master corporal in 444 Squadron. The Brigade had been "snowballed" after the Russians had massed on the Polish border and every one was being deployed to what was then known as the Real Survival Area – not an overly reassuring title for a forward deployment area. I joked with them that at that time I had thought my chosen profession was not that sensible and that driving alone at night towards the East German border and the Fulda Gap in a five ton truck full of JP4 was a very frightening experience. I finished off by telling them not to bury the fear but to acknowledge it and then work through it. I assured them that their training would kick in when it was needed.

Fatigue

Fatigue became a huge concern during the deployment. The aircrew were maxing out on their crew days and a good night's rest was impossible. The support personnel were running on coffee, trying to maintain a long supply chain and preparing to move forward. The night before the advance party's departure into Kosovo, no advanced party personnel were placed on guard or kitchen duties, and all drivers were ordered to bed for a forced rest of at least six hours. Although the move was only approximately 250 kilometres, it took almost the entire day as the route into Kosovo from Skopje was bumper to bumper.

We arrived at our in-country position near the town of Glogovac, Serbia in the late afternoon. We set up a small security perimeter and posted guards. At dusk the first of the squadron's sea containers started to arrive. They continued to arrive throughout the night.

At one point I was talking to a corporal (Cpl) driver of the sea container transport vehicles. It was obvious he was basically dead on his feet. I asked him how long he had been up and he indicated it had been over forty-eight hours. Although not part of my unit, I ordered him to get four hours sleep in his truck before I would release him.

One of the greatest follies of being in a leadership position is that you will work yourself longer and harder than what you would expect from anyone one else. In my case, I had been working a minimum of twenty-hour days for about three weeks. What finally caught my attention about my own personal fatigue level was a series of small personal accidents. In the space of four days, I had cut my hand twice, banged my head three times, once requiring medical attention, and had decided to flatten my thumb with a five pound sledge hammer while setting up concertina wire on the camp perimeter. My body was taking punishment for not getting the rest I needed. I shortened my workdays to eighteen hours and found I was able to maintain that work pace indefinitely.

Fitness

We in the Air Force pay lip service to fitness. We continually find excuses to avoid implementing a unit fitness plan. During my deployment to Kosovo it was self-evident that those members of the unit that had a fitness programme back in Canada were better equipped to endure the long hours in forty degree Celsius heat. I have been a long time believer in physical fitness, and even though it may seem like a poor waste of sea container space, we did bring a small amount of fitness equipment with us. Once we had settled into a steady routine, everyone in camp was encouraged to maintain or improve his or her fitness. The gym was in use twenty-four hours a day, seven days a week.

Welfare

Even in high temp operations, the airmen and airwomen need an emotional outlet. Although the work never stopped, the unit was encouraged, when possible, to have one down day a week. This rest was coordinated by the flight WOs. After the first month, the rest and relaxation (R&R) trips started, as did special event days. In our case, our first day to unwind was Canada Day. The unit went to a minimum manning posture and sporting events were organized. With a little help from some embedded British signals personnel, we were able to get a unit photo taken. The day finished with a barbeque and a movie in the mess tent. As the movie went over so well, it became a nightly ritual. Other ways to reduce stress were also brought on board as the tour progressed, including: foose-ball tables, electronic games and board games.

Once the camp was set up and established, communications systems were set up and a Local Area Network (LAN) and telephone system was put in place. Later, a rear area satellite link was established with Canada and the DIN. All camp members were now able to phone home on a regular basis and/or to send and receive e-mails through the two computers that were made available for that purpose.

Traditionally, at around month four the morale of a tour takes a big dip. We avoided this drop in morale. At the four month mark, the camp moved out of modular tentage and into what is called "Mech Shelters," complete with individual privacy screens. New state-of-the-art washrooms and shower facilities were opened and, in terms of welfare programmes, the Canadian camp became the envy of most other NATO counties.

Cultural Concerns

The KRWAU was a mixed-gender and multi-denominational unit which led to some unique challenges. The unit had two corporals who were living together as a common-law couple back in Canada. Prior to our disembarkation, the whole unit had been briefed on the standard of professional conduct with regards to fraternization. This case was a perfect example of two individuals working in a close relationship in the same unit. In every respect, they conducted themselves as no more than good friends. Unfortunately, not everyone agreed that this was the case and I received complaints about their eating meals and watching a movie together.

I called the complainant in and we discussed the member's concerns. The member admitted that no open fraternization had been observed but that he suspected that there had been a closer encounter as was obvious, according to him, from their spending so much time together. I asked if that criteria – spending time with someone – was used to determine if others in the camp were engaged in fraternization. I gave other example of unit members who ate meals together and spent time together in the evening. I included the CO and myself as an example of two individuals who spent an enormous amount of time together. The members said we were of the same sex and that made it different. I explained that that was the exact point and that all unit members regardless of gender would be treated equally.

The KRWAU also had one member of the Muslim faith. He came to me early on in the tour and we discussed his religious requirements. As a signalman, he was required to work shifts but, with concurrence of the signals sergeant, every effort would be made to allow the member to meet his daily prayer requirements. I also had the Chief Cook and his staff prepare special meals for the member. The Muslim holidays of Ramadan provided another challenge, as the meal hours fell during daylight hours when the member was not allowed to consume any food or drink. Once again, the Chief Cook accommodated and made up late meals for the member so he could eat after sunset.

Rules of Engagement

During our initial entry into Kosovo our Rules of Engagement (RoE) were very robust, with almost everything considered mission essential and therefore affording the use of deadly force in a wide variety of circumstances. As the operation progressed, the RoE became more stringent. Updated RoE briefs were a frequent requirement. Changes in the RoE were briefed through the chain of command. After every change to the RoE, I would wait a few days, then walk the camp questioning individuals on these changes, thereby ensuring the information had been passed throughout the camp.

Leadership

On a dynamic forward operation, the demand on the leaders and their leadership is enormous. Strict discipline and compliance to standing operating procedures (SOPs) must be adhered to. This compliance is essential. Our unit had an incident during which some member of the camp had been observed to be impaired and, therefore, in non-compliance with the SOP of two beers a day. An investigation ensued. During the investigation it was determined that a sergeant had been present when the consumption of alcohol in excess of two beers had taken place, although he himself had not partaken. Regardless, I laid charges against this sergeant for not doing his duties and enforcing the camp standing order in regards to alcohol consumption. The sergeant elected a court martial. His court date was months away. I now had a Sr. NCO in camp whose confidence as a leader I had lost.

After a long discussion with the CO, the sergeant was repatriated back to Canada. This may seem like a punishment before the trial, but it is the prerogative of a deployed CO to send anyone back to Canada if he or she has lost confidence in that member's abilities to perform their duties. In this case, both the CO and I were in perfect agreement. As a sergeant, the member did not have the right to choose what orders he would enforce. Although this incident occurred within the last two months of the deployment, it did produce some positive results. The Sr. NCOs immediately took on a more proactive role in the application of all their duties. This is not to imply that members walked around camp scared they would be fired out of theatre for any small indiscretion, but they now understood that the cold face of discipline resided in their hands and personal leaderships was something that was always required.

COMPOSITE UNIT

The Mission

For my second tour, I volunteered for Operation Apollo. I had initially been selected to be the Tactical Airlift Detachment (TAL Det) CWO for the C-130 Hercules aircraft, supporting the war on terrorisms at Camp Mirage in the South West Asia (SWA), Area of Responsibility (AOR).

About two weeks prior to my departure, I received a call from the Chief of the Air Staff CWO who informed me that shortly after my arrival I would become the camp CWO. The whole command and control structure of Camp Mirage was to be changed and our mission would change from Operation Apollo to Operation Athena. Our new mission was to provide both a logistical and air mobility support to Task Force Kabul (TFK) in Afghanistan.

It needs to be stated that Camp Mirage was a mature facility which had been in place for over eighteen months. The vast majority of the camp was hardened structures.

Communication

For the most part, there were relatively few communication difficulties. The rear link to Canada had been established and had been functioning well for years. There was more than enough computer access for the camp members to e-mail, message or make telephone calls home. The only problem in this area came when two members in Kabul were killed when their jeep struck a land mine. It is standing operational procedure to close all communications back to Canada when a significant incident occurs. This Lock Down ensures that the next of kin in Canada are notified by the proper personnel with emotional support people in place and not by over the fence gossip by a well-meaning neighbour who may have just got off the phone with their spouse back on the deployment.

A Lock Down occurs without notice and even the camp commander may not know the reason why. As this was our first Lock Down, most personnel in camp were quite upset that their lines went dead or their computer screens blank. The solution to this was simple: it was fundamentally a matter of educating the camp on the need for a Lock Down policy.

This experience led us to another communication tool. We set up two information screens in the dining hall. These screens ran a power point show on a continuous replay loop. All general camp information was passed via this means and was controlled and coordinated though my office. I also instituted a weekly Sr. NCOs and WOs coffee break and attended weekly meetings with my coalition counterparts from the Australian, New Zealand and American air forces.

Team Building

Camp Mirage consisted of 278 members from across Canada. With the exception of the aircrew and the Defence and Security platoon, no other section had all its personnel from the same base in Canada, let alone from the same unit. As our operation and mission had changed, so too had our command structure. We were now part of TFK and under direct authority of the Task Force Commander in Kabul. Team building thus needed to be done on three levels: the Task Force level; the Unit level; and the Section level.

For the Task Force level, team building was accomplished by a series of briefings to all the TSE members on the Task Force mission in Afghanistan and where we fit into the larger picture. A new camp crest was designed along with a new camp motto: "One team, One Mission." This was an outstanding choice because it related to all levels of the team building and sent a message to the Task Force in Kabul that we were part of the mission, regardless of our geographical location or the colour of our berets.

The section level team building was more complex. The sections only had a couple of weeks to become a well-oiled machine, before more than twenty-five hundred soldiers transited through on their way to Kabul. We did not have the luxuries of a pre-deployment exercise or beer calls. All the members had in common with their section was the oppressive forty-nine degree Celsius heat and the 100 per cent humidity, coupled with a mind numbing amount of work to prepare the camp for intense operations. For example, the supply section had over 1000 sleeping bags stacked in a corner and the orderly room had 57 bags of dead Operation Apollo files to go through and either file, destroy, or return to Canada.

To repeat, nothing builds teams like sharing hardship, and for the first two weeks everyone was on an eighteen to twenty hour workday. The sections quickly became effective teams, largely due to the leadership of the Sr. NCO cadre.

The unit team building took the longest to achieve and was very hard to implement. This difficulty was largely due to the camp layout. There were only two common social gathering areas for the entire camp. The first being the dining hall and the second being a mess of sorts that was one kilometre up the road from the main camp, housed in a poorly air-conditioned building. The camp consisted of 14 barrack blocks. In front of each was a small sun screened patio. Rooms and barrack block allocation was done basically by sections. As no one used the mess due to its distant location and because people tended to eat with their peers, the camp had 14 individual teams working in isolation from one another. Very little cross section interaction was taking place.

I approached the CO on this matter. I informed him that there are monies available on tours to do quality of life projects and I recommended that a large common sun screened patio be constructed next to the dining hall. This idea was accepted and implemented.

Special events were then organized including a campaign to name each barrack, bingos, evening movies played, sport days and a unit photo. These tools all contributed to bringing the whole camp together.

When performing team building, it is imperative that everyone on the team is remembered and included. At Camp Mirage we had eight civilian employees who were assigned to the camp from the Canadian Forces Personal Support Agency (CFPSA). I included them in all aspects of the camp. Their site manager was treated as any other Flight Warrant Officer and their staff was made to feel a part of the military family. During special events, they were always included. For example, a member of their section was selected to do one of the readings for Remembrance Day and for the memorial we held in camp for the two TFK members killed by the land mine.

Mission Change

Camp Mirage had originally been set up to support air operations for Operation Apollo. One month after our tour began, our mission changed to provide air mobility and logistical support to TFK in Afghanistan. We were no longer the focal point of an operation in the AOR. We were now a support element of the new Canadian operation in Kabul, Afghanistan.

With the exception of equipment, all support flights in and out of Afghanistan originated from Camp Mirage. Due to the threat assessment, the only approved method for personnel to travel to and from Kabul was via a C-130 Hercules aircraft from Camp Mirage. It soon became obvious that our mission was essential to TFK. In the words of the TFK contingent commander, "Camp Mirage is critical to Op Athena. I considered you guys the defenders of the pass. If camp Mirage fails, quite frankly the mission fails and soldiers will die."

This new focus and mission was passed down to the airmen and airwomen through the chain of command. Shortly before Operation Athena commenced, I did my first of several visits to Kabul and Camp Julian. I took the time to take as many digital pictures of the operation in Kabul as possible, including quite of few pictures of the soldiers doing their work. When I returned to camp these pictures were used to reinforce our role in this new operation. Additionally, when a seat was available on one of our daily flights to and from Kabul, TSE members were encouraged to fly to Kabul to see for themselves why their job was so important. Through these efforts the camp quickly switched over to their new role and mission.

Morale

Shortly after arriving it became apparent that the camp required a major change in its attitude especially in regards to dress and deportment. Before I go into these observations, let me state that this in no way is a reflection of the previous camp administration. They had been on a different operation. Op Athena, Roto 0, which the camp was now on, required a very different approach.

The tactics I used may seem draconian but one should bear in mind that my time was very limited. In just a few short days, Camp Mirage would be overrun with over twenty-five hundred soldiers transiting through on their way to Kabul and the camp needed to be running at peak efficiency.

I started by having all-available personnel, approximately 230 people, perform a Foreign Object Damage (FOD) walk down the flight line. This procedure was done twice. Then the camp work site was policed for trash twice; a check of the accommodation site followed and this too was repeated. Once I was satisfied with the tidiness of the camp, I had everyone report to the dining hall. I then explained my great displeasure of the camp cleanliness – or lack there of. That was the last time that we needed to perform a camp-wide trash sweep. I went on to explain that the amount of work that lay before us would not allow the camp to have each of the 230 personnel away from their jobs for two and a half hours in order to do camp cleanliness sweeps. Even if the camp did cleanliness sweeps once a month, we simply could not afford the lost 575-man hour a month for each sweep.

I informed everyone that camp cleanliness was everyone's job, all the time. No exceptions would be tolerated. Smoking would be restricted to designated areas only. Lastly, I informed the camp of the new dress policy. Until then members had been allowed to move around the complete camp in T-shirts. No name or rank identifiers were visible. Effective immediately combat shirt were required in all common areas and hats were mandatory. Members could strip down to T-shirts in their designated work area only. When in T-shirts, a rank slip would be worn on the belt just left of the buckle. The use of first names would no longer be tolerated.

So how does all this dress, deportment and discipline relate to morale? Quite simply, previous to these changes, the camp had no esprit de corps. It had no military pride in itself or as a members of the CF. The camp was little better than a collection of civilian employees in green clothing. I could discern no military foundation to use as a compass to guide it though the long hours and tough times that I knew lay ahead.

This stricter approached initially produced the expected huge drop in morale and was not well received. The camp Sr. NCOs and WOs came to me in a wave. Their biggest concern was the wearing of combat shirts. Did I not know it was over forty-five degrees Celsius outside?

I informed them that I did and that I had taken that into consideration. I was unwilling to bend on this item, however. Camp Mirage consisted of three coalition units. The Australians, the New Zealanders and us, the Canadians. The Australians were in full shirt now and always had been, the new Squadron WO for the New Zealand contingent was in agreement with me and his personnel were being informed of the new policies. Additionally, the camp was located on a host nation air base. Consequently, moving to and from the work site and accommodation area required the camp members to be visible to the host nation and also the American air force air transport wing stationed further down the flight line.

I then informed the Sr. NCOs and WOs that the new policies were in place and I was not going to budge on them. Furthermore, I would not be the only dog that barked in camp and I demanded that they enforce these orders. Those individuals who were unwilling to do so would be accommodated with a one-way ticket back to Canada. Perhaps this treatment was harsh. Nonetheless, it was effective.

It took almost two weeks before I saw the positive results emerge in the morale of the personnel. The TSE were now outwardly displaying a new found confidence, panache and bravado. They now demonstrated pride in themselves, in their unit and in the CF. Their professionalism shone throughout the camp. This new positive attitude was mentioned to me several times by very important persons (VIPs) transiting through Camp Mirage. They had visited the camp before and noticed the difference. My peers in the coalition units also saw an improvement.

One of the benefits of being a rear area unit is that you generally have more freedom and better accommodations than the forward units. In our case, the camp had walking out privileges to a major city in the region. Daily bus trips facilitated these excursions.

Early during our tour when the members of TFK were transiting through Camp Mirage on their Home Leave Travel Allowance (HLTA), it had been decided that the members from TFK would be restricted to camp for their twenty-four hour or more hold over in Camp Mirage. I felt it was unfair to impose this restriction on twenty-five hundred soldiers for the discrepancies of perhaps a few. I discussed this with the Deputy Commanding Officer (DCO) and during a CO's executive meeting we approached the

CO recommending that the TFK members be allowed to go downtown and enjoy the city like the TSE members. The CO was sympathetic but initially did not budge in his position. The restrictions originated because thirteen members from the Kabul theatre activation team had abused the walking out privileges and remained out past curfew earlier in the tour causing a huge amount of work and security difficulties had ensued.

Prudence would have had me drop the subject then and there, as I had learned that once my CO had made up his mind and came to a decision, he expected it carried out without debate. I felt this issue was important enough to press, however, and spent a couple of hours with him discussing the topic. In the end, he conceded to our proposal. The condition was that each HLTA flight coming out of Kabul would have to receive a thorough in-brief prior to being released to go downtown. Of the 100 HLTA flights required to transit the twenty-five hundred TFK members out of Kabul for their leave, I personally briefed 75 of these flights. I felt it was a small price to pay to ensure that everyone was being treated equally.

Fatigue

Although Camp Mirage was well set up and the accommodations were excellent for a deployed operation, fatigue was initially a problem. Due to the change in operations, all unit standing orders had to be rewritten and they had to mirror those of the TFK. We had to receive, feed, house and provide all the other assistance required by the over twenty-five hundred soldiers transitioning through. On top of this requirement, we had to perform the mandatory close out Board of Inquiry for a mission that had just ended. It was a staggering amount of work and it all had to be completed in the first thirty days. Once again, eighteen to twenty hour work days for everyone were the norm, in blistering heat and humidity. We had daily cases of heat prostration in the camp hospital. Section supervisors were briefed on the symptoms of fatigue and heat prostration. Absolutely no loud noises were tolerated in the barrack blocks or even on the accommodation side of camp. The supervisors did their best to monitor their personnel during this surge of work.

As for myself, I once again found out the hard way that I was working too hard. Because of the long hours I had enforced upon myself, I had depleted my immune system and contracted food poisoning. Additionally, the joke of the camp was that my office was the third stall on the left. Even then I continued to show up and do my job. It was not until three and a half months into the tour that I finally took a complete day off, as ordered by the CO. As I look back, I see several areas of responsibility and decisions in my job that I could have done better had I been properly rested and in good health.

Lastly, in regard to working hours, Camp Mirage had over the years picked up an undeserved reputation as "Club Mirage," a great place to go for a six month tour in the sun and sand. With 278 members supporting over twenty-five hundred soldiers, the support tail on this big operational dog was very small. There is no way TFK could receive its required support on an eight hour day, five days a week with beach parties every night work schedule. The camp put in hours equivalent to any unit in a forward area and in some of the harshest weather conditions imaginable.

Fitness

Once again the Air Force attitude to fitness became an issue. In fact, compared to our coalition partners, we were on the whole overweight and out of shape. One only had to look around the camp to question the fact that everyone had passed the CF Express test for fitness, a mandatory requirement prior to deployment. It was my observation that those members who were overweight and out of shape were the ones reporting to the camp medical section with heat prostration problems. There were of course exceptions whereby very fit personal had overworked themselves in the extreme heat, but generally speaking the problem resided with unfit and/or overweight members.

The CO approached me about this issue and we had a long discussion. Our views on fitness were the same and he felt (and I concurred) that a complete assessment of the camp's fitness level needed to be conducted. Every member of the camp was required to report to the camp fitness coordinator from CFPSA for a fitness evaluation. As a qualified fitness expert, she would perform a short interview with the individual and between a series of measurements and a review of the members pre-deployment express test sheet make a determination of the member's general fitness. In the end, it was her expert opinion that over 60 per cent of the camp was in poor physical fitness.

This assessment also produced another interesting result. A large number of the camp members had no documentation of having passed the express test in the six month time period prior to deployment. For the most part, their home units were also unable to confirm if their express test was ever done.

Based upon this information, the CO initiated a fitness plan for the camp. All members were encouraged to improve their fitness and every month a camp fitness walk was conducted by the entire camp.

One of the major points we had to drive home to all camp members was that, as part of TKF, the contingent commander could order anyone forward as required. The minimum fitness level for deployment to Kabul was the Army's Battle Fitness Test. Once a member moved forward they had to wear the complete flack and ballistic vest, helmet and carry their weapon at all time – meaning that they would be wearing a 60 pound load in downtown Kabul.

Importantly, the CO and the rest of the senior leadership set the example on what was expected. The CO, who ran daily, invited anyone from camp to join him. The operations officer and myself could be observed riding bikes around the camp. The deputy commanding officers lead evening walks for those who did not run. This demonstrated leadership produced the results we hoped for and, finally, the vast majority of camp members initiated their own fitness plans.

Welfare

The welfare programme in Camp Mirage was outstanding. Shortly after I arrived, the welfare coordinator's position, which had been lead by a Navy Chief Petty Officer Second Class, was filled by civilians from CFPSA. Their job was to coordinate the R&R and HLTA for Camp Mirage personnel and provide welfare support to TFK members transiting through. They also ran a small exchange and kit shop and, through their fitness coordinator, organized sporting events. The CFPSA personnel were a handy tool at my disposal. Because of their duties and responsibilities, they always had a good sense of the morale of the camp. Their civilian status made most members at ease and they had no qualms about telling them what they liked or disliked about their tour. I would drop by the CFPSA section daily and chat with the site manager or another CFPSA representative. I would ask if there were any problems that had surfaced either with Camp Mirage personnel or with soldiers from Kabul transiting through. If there were, I would subsequently investigate the problem and implement a solution to ensure it did not reoccur.

The camp was also blessed with an outstanding padre. When the two soldiers died from the mine strike, she took the lead and coordinated all Camp Mirage activities in this area. She approached me and asked if we could hold our memorial service until the evening. She reasoned that one soldier arriving in camp from Kabul on the daily HLTA flight was from the same section as the two soldiers who had died and because of flight schedules he would not be able to attend Camp Julian's memorial service. This type of attention to detail in regards to the welfare of not only Camp Mirage personnel but also all TFK personnel was present through the Camp Mirage leadership.

Cultural Concerns

Our mission was simple yet our geographical location in the Persian Gulf area did produce a few cultural issues. Our location in an Islamic country conflicted with many of our western ethics and values. For example, public drunkenness was a criminal offence, as was drinking and driving. Notably, the deployment did not have a Status of Force Agreement (SOFA) with the host nation. This meant that should any of our airmen or airwomen be arrested for public drunkenness, (something we would consider a minor offence), it could result in incarceration, flogging and/or deportation. This punishment could occur from something as simple as a taxi driver smelling alcohol on a member's breath. We all fell under the laws of the host nation. Moreover, they were not even obliged to notify us of the arrest of our personnel.

A large portion of our camp-in brief was devoted to the host nation's laws and cultural differences. To ensure everyone was aware of these differences, all personnel were restricted to camp until they had received a proper briefing. This included all transits, counting VIPs.

Having come across this problem on my first tour, it was obvious to me that the Islamic celebration of Ramadan would require some extra cultural awareness training. Although not a huge problem on camp, the consumption of any food or liquids during daylight hours was not acceptable behaviour off base. A whole series of lecture and information slides were distributed. The camp padre took the lead on this issue as well.

Rules of Engagement

Camp Mirage was on the lowest readiness and alert level. Our rules of engagement (RoEs) were quite simple: detain any suspect with minimum force until the military police arrive. Deadly force was only authorized when a member felt his or her life was in immediate danger.

Many of the camp members were required to travel to Kabul. For those of us who did so, we were required to know and understand the RoEs for Kabul. Those camp personnel travelling to Kabul would receive a TFK RoE brief prior to departure.

Leadership

The living and recreational facilities at Camp Mirage were far superior to any other Canadian Camp in the AOR. This high standard had a tendency to cause some members to allow their leadership responsibilities to relax. On one occasion a senior supervisor in the camp was loudly griping to her superior in the dining hall about having worked 11 days straight without a day off to go downtown. The member's superior was trying, with limited success, to get the member to lower her voice. Instead the member became more agitated and several tables could overhear the conversation.

At that point I approached the member and asked her to step outside as I had something important to discuss. Once outside and away from prying ears I quietly informed the member that she was to stop whining immediately and that public whining by a person of her rank was very inappropriate. I explained that at the next table over were the two camp linemen who had not had a day off in the three months they had been on camp. I also explained that these two corporals had overheard the whining and that the member's stature had been diminished in the corporals' eyes. When the member started to argue her point, I informed her that this was not open for debate and whether she liked it or not she would cease the whining immediately. As tempers were up, I requested the member drop by my office in the morning so that we could discuss this issue in more detail. The next day the member freely admitted that she was out of line and should not have acted as she had.

This incident led me to reassess the open dining hall policy and contemplate whether segregation by rank was required in the dining hall. I pondered this idea for several days taking into considering the layout of the dining hall, our coalition partners on camp who shared the facilities, our mission, the fact that a lot of camp business was conducted over meals that required multiple ranks and the overall effect it would have on unit morale. In the end, I decided against the change. Instead, I kept a closer watch for incidents of this nature.

Concluding Remarks

As you can see, regardless of the type of deployment, formed or composite unit, the leadership challenges are the same and require tough moral choices and general adaptability. How a leader applies appropriate leadership is dependent upon the situation and conditions at hand. The one point that I would drive home is that the closer you are to the Forward Edge of Battle, then the greater the requirement for discipline, both institutional and self-discipline.

My first CO, Lieutenant-Colonel Herm Harrison, said to me when I became 450 Tactical Helicopter Squadron Warrant Officer, "The first duty of leadership is to lead; the first order of a leader is: Follow Me!"

CHAPTER 9

SETTING STANDARDS FOR THE FUTURE: ADVICE FROM AN RSM

André Jutras

From February to August 2004, I acted as the Regimental Sergeant-Major (RSM) of 5 Canadian Service Battalion, during our deployment to Afghanistan on Operation (Op) Athena, Rotation (Roto) Zero. My role was to maintain discipline and morale and assure the well-being of the troops for the Commanding Officer (CO).

These responsibilities applied to military personnel and civilians alike. Under my charge, I had 327 military personnel and 257 civilians from the Canadian Forces Contractor Augmentation Program (CANCAP), as well as 51 civilians who were from the Canadian Forces Personnel Support Agency (CFPSA). These members came from across Canada, which made the task very enriching, but also very demanding.

Of course, I did not do this work alone. Each group had their own individual in command (IC), who had to make sure that everyone, without exception, complied with the camp discipline and law, and had up-to-date information on the situation in terms of dangers they might be faced with.

With the exception of military groups such as mobile support equipment (MSE) operators and movements and ammunition technicians – totalling roughly 80 to 100 people – the entire group worked mainly in the camp and did not have to leave the wire on a regular basis. Nonetheless, we had to make sure that everyone knew what to do should an emergency occur. On our rotation, the danger level was fairly high, so we had to maintain a high alert level in order to be ready to respond quickly to any occurrence. For most of the people on site, whether they had to go out of the camp several times a day or always stayed in the camp, not knowing whether we could deal with hostile action had a deep psychological impact. We had to prepare the personnel with routine alert exercises, awareness readings and many words of caution throughout the mission.

The National Support Element's (NSE's) responsibilities were many. Basically, we had to provide the logistic support for the Canadian Battle Group everywhere in the theatre. This was no small tasking. Duties ranged from maintaining camp infrastructure, providing travel service, maintaining sports facilities, meeting personnel's essential needs in the camps, moving equipment and personnel to airports, and taking care of entertainment in messes and canteens, to name only a few. We not only had to maintain morale among NSE personnel but, mainly, we had to provide the battle group with all the support possible in order to make their task easier and to keep their morale as high as possible. On many occasions, NSE personnel had to act selflessly in order to give the best possible service to the battle group.

Our mission was a resounding success. Without the leadership shown by supervisors and troops, however, the mission could have encountered serious problems. I must admit that luck was often on our side but I am equally convinced that the close supervision conducted on this mission greatly contributed to its success. It is not easy to have so many people work effectively together, particularly when they have such varying degrees of understanding as to what a hostile environment can be.

Our personnel, particularly those who had to leave camp, had to be kept abreast of the enemy situation at all times. We also tried to preserve high morale while keeping the focus on the success of the mission. We had to keep personnel busy but, at the same time, we had to be very careful not to tire them out uselessly. Each supervisor was well informed as to his or her responsibility and had to be as transparent as possible to the troops. I made sure I met with platoon second-in-commands (2ICs) to pass on to them the Commanding Officer's (CO's) vision and to resolve any problems that might arise.

In a theatre of operations, it is impossible to leave requests from the group unanswered. This did not necessarily mean that we approved every request from the troops but, at the very least, an answer had to be given. I believe this greatly contributed to keeping up the troops' level of satisfaction and also provided a level of confidence in the chain of command.

The CO had given his platoon commanders a fairly broad guideline so that they could make decisions without always waiting for his approval. This also gave platoon 2ICs expanded responsibilities without taking away the platoon commander's final decision-making powers. Thus, platoon 2ICs could provide information to the platoon commanders so that they could make informed decisions. I think this way of doing things was well received by platoon commanders and 2ICs. It was clear that the chain of command supported them.

In return, the CO requested a weekly meeting with his platoon commanders in order to receive a detailed summary of each platoon's accomplishments, their problems and anything else that seemed relevant. Obviously, any platoon commander's or 2IC's emergencies had to be dealt with immediately. For my part, I met with all platoon 2ICs every week for an update on routine tasks, directives, clothing and disciplinary matters. I held this kind of forum to make sure that everyone had the same command philosophy, that everyone was working in the same direction and that the CO's directive was understood and followed.

The same diligence was applied to civilian personnel. One disciplinary authority per civilian group came to my meetings and took notes. In camp, there were no separate policies for civilian and military personnel. If it was zero tolerance, then it was zero tolerance for everyone.

One of the major challenges that we had to overcome was instilling a sense of urgency in civilian personnel during danger alert practices. The fact that they were not exposed to the hostile reality outside the camp as much as their military counterparts were meant that the real danger of rocket attacks on the camp was somewhat unfathomable to them. Exercises, unfortunately, were not enough to make them feel a sense of urgency. The CO had to be very firm and persuasive in order for the civilian personnel to understand the need to be always on the alert and achieving this state of readiness among the civilian personnel remained a constant challenge throughout the entire tour.

Among military personnel, there was a constant risk of terrorist attacks outside the camp and the fact that we had to cross a city of 2 million people in very heavy traffic, often getting stuck in traffic jams and slowing to a standstill for several minutes at a time while not knowing if someone near us was hostile, was an extremely difficult experience for all. We applied psychological tools to relieve the tension in the group. Personnel were required to attend briefings that encouraged them to share their fears and anxieties and to talk about their personal experiences outside of the wire so that their awareness of the situation could benefit others. These discussions were very helpful as they provided many individuals with reference points in case they ever found themselves in similar situations. Moreover, it helped some members put bad experiences behind them for good, while providing others with a response capability.

I made a point of attending these briefings so that personnel would know that I sympathized with their concerns. It is important that the chain of command gets involved but, most of all, it is vital that troops know that the chain of command is available to them. Everyone, without exception, was feeling tension either from work or from being far away from family members, so we had to be attentive.

Throughout both the training and the mission, I made an effort to exert leadership that was firm but fair. All disciplinary cases were investigated and if it turned out that charges had to be laid, then I was always the one who laid them so that there would be only one standard and people would know and understand their left and right arcs.

On the other hand, I tried not to interfere in other people's areas of responsibility. Platoon 2ICs and their supervisors must be able to exercise their leadership in such a way as to avoid being taken advantage of. Distance supervision is necessary in order to ensure that the focus of the mission is maintained and that military values are upheld. While there is absolutely no point in smothering our supervisors with questions and extra work, neither should one be unaware of what is going on in the unit.

Obviously, we had to face a number of challenges throughout both the training and the mission. Indeed, forming a unit whose personnel came from several different locations and environments and to succeed in turning it into a unit that was trained, disciplined, and ready to leave on a mission so perilous as the one in Afghanistan was no small feat. We had to have training that was imaginative, interesting and, most of all, which accurately targeted the tasks we would have to do in the theatre. This training had to be done with a constant focus on the safety of our personnel in a type of threat situation to which we were not very familiar: terrorism.

In hindsight, I think we did very well on the mission. There were no casualties and no major accidents among our personnel. The training was beneficial, the supervisors in theatre all did their work in an exemplary fashion and NSE personnel managed very well.

Of course, if I could have changed things, I would definitely have liked to have had a better understanding of the civilian contractors working on the camps. While civilian contractors undeniably made a significant contribution to the success of the mission, a better understanding of them would have made us much better prepared to use a more effective approach in making them aware of the effort they had to put forth in order to ensure their own safety.

I still feel uncomfortable describing the exact responses of our civilian workers with regard to safety concerns. Personally, I do not think that these people properly understood what kind of potential danger they were in. One could sense a type of carelessness on their part which was even observable when we practiced safety scenarios. The camp security plan was excellent, however, the speed with which personnel took cover in case of attack was vital in order to minimize losses. The response among military personnel was good but the civilian response was much less so. Consequently, military personnel would have had to protect the civilians in a direct attack and I am convinced that this would not have been easy. Finding a foolproof solution to this problem is no easy task but better preparation of the civilian people hired by the contractors and a deeper awareness among those personnel of the danger inherent in this type of work would have been beneficial.

Another aspect on which I would have placed more emphasis, and with which current military knowledge still remains quite vague, is on the understanding of military convoy movements in an austere urban environment. If we look at the problems the US Army in Iraq must face every day, it is clear that this topic must be taught in much more depth in order to better prepare our troops for this type of conflict.

Since no two deployments are exactly alike, I do not have any specific advice to give to the people who will succeed me on this mission. Nonetheless, I can provide some general comments.

The success of the mission depends on the people serving on it and they need to be given as many opportunities as possible beforehand to work together in a training environment and to feel comfortable and supported by the chain of command. For this type of preparation to occur, the concept of training must be changed to make it more realistic and as close as possible to the harsh reality of today's missions as possible. If we want to minimize our losses, then soldiers have to know what to do in any contingency. This is currently not the case. A new doctrine is being drafted but, in the meantime, we are already training soldiers for missions that will be held in the very near future. It is absolutely essential that all Canadian Forces (CF) personnel be immediately exposed to this new type of conflict training. It would be wrong to think that because some individuals are in a support occupation, they are not exposed to hazards.

In closing, I think that the principles of leadership applied during my tour can be best described under the dictum "an iron hand in a velvet glove." Mutual respect among military personnel, civilians and people of other nationalities working and living in the same camp requires several ongoing adjustments so as not to lose one's sense of values while still staying focused on the mission and its objectives. If the work is done with respect and everyone feels that he or she is contributing a little something to the community, then we are definitely on the right track.

It is certainly not easy to please everyone, especially when everyone is under a certain amount of stress during their tour, but to try as much as possible to give people what they ask for so that they know that the effort is there is a key to success. Believe me – and I say this humbly – the success of our mission largely depended on the leadership that each individual person exercised. The mission was not the responsibility of one or two people; rather, it was truly a collective effort. Certainly, man is almost guaranteed success when individual effort is made as part of a collective whole.

Only very rarely has Canada had to face a terrorist attack and I know we were not prepared for such a war as the one currently ongoing in Afghanistan. I do not think that we should change our way of seeing things simply because we are on a mission. There has to be leadership on the ground and we need to make every effort to get to know our personnel well and, especially, to lend them an attentive ear. It should never be forgotten that on these types of missions, everyday men and women risk their lives. It would therefore be very wrong of us not to listen to them. Once again, we cannot grant their every request but a sympathetic ear and a good explanation can alleviate a lot of frustration. Mutual trust in a team is essential. When people join the group late, with only two weeks of pre-deployment courses, however, it is very frustrating for everyone who has had to do the full training, often far from family members. I know that these people are the exception but it is nonetheless frustrating and such personnel have tremendous difficulty adjusting to the larger group.

Let us make sure that we have enough personnel in the various occupational groups to meet the demands of ever more frequent missions with support occupations. Currently, this is not the case. If Land Force receives a certain increase in strength then support occupations will also have to receive a substantial injection of new blood in order to effectively meet the demands of missions to come.

CHAPTER 10

ESTABLISHING A CANADIAN FOOTPRINT IN AFGHANISTAN: INSIGHT FROM AN RSM DURING OPERATION ATHENA

Michael Poirier

I was the Regimental Sergeant-Major (RSM) on Operation (Op) Athena. We were the first rotation, sent in to replace the 3rd Battalion, Royal Canadian Regiment (RCR), which had established the mission.

The objective of the mission was to assist the Afghan Transitional Authority (ATA) in providing a safe and secure environment in western Kabul. This required us to conduct several patrols, both night and day, mounted and, especially, dismounted. Foot patrols allowed us to have much closer and more direct contact with the population and, through local interpreters, to engage them in conversation. Consequently, the local population perceived us as soldiers who were concerned with their problems and challenges, rather than as an occupying force. I feel that this is a very important distinction. I am convinced that this approach and our concern for the Afghan population made a real difference and contributed to our being able to bring all our personnel safely back home to Canada.

As stated, the goal of our operation was to make our presence felt on the ground through mounted and foot patrols. We also planned a number of area saturation operations at the section, platoon, and company level and we even had one at the Battalion Group level. These operations were necessary in order to send a clear message that we controlled our area of responsibility. We conducted several roadblocks at irregular intervals to control the trafficking of narcotics, explosives and weapons. We also planned and conducted direct operations that I am not at liberty to discuss in detail.

In contrast to the Balkan theatre of the early 1990s, no belligerent forces threatened us or local organizations. However, there were local criminal gangs and organizations that jeopardized the formation of viable democratic and economic institutions, both at the local and national levels. We were under the constant threat of suicide attacks, booby traps and car bombs, and always received a security briefing and a situation report (sitrep) before leaving Camp Julien. Moreover, we needed to be doubly vigilant during our patrols to ensure a maximum level of security. Near the end of our mission, we were granted national authority to make our presence felt in remote areas outside of Kabul as part of the Provincial Reconstruction Team (PRT).

During this mission I was the Regiment Sergeant-Major (RSM) for the 3rd Battalion, Royal 22e Régiment, Battle Group (3 R22eR BG), which included an engineer squadron. My main responsibilities were the "three Ds": drill, dress, and deportment. I exercised these responsibilities both inside and outside Camp Julien. I was also responsible for the morale and well-being of the troops, a task made easier by the quality of the camp: it was the highest quality camp I have ever seen in my entire career, in all my tours of duty. I also accompanied my Commanding Officer (CO) during various inspections on the ground and during certain patrols with the troops.

My greatest challenge was to ensure that the troops maintained the highest standards of dress, discipline and deportment at Camp Julien. Given the variety of Canadian and international units at the camp (including Americans and Slovenians), we had to ensure that the troops maintained a high standard while allowing them to relax when they returned from a mission or an operation.

There were nearly two thousand personnel in this very large camp, divided into various groups, including the Canadian Contingent's group of senior staff (National Command Element), a service support group (National Support Element), and the various components of the Task Force, including the Battle Group and about 350 civilian employees of several different nationalities. We therefore had to learn to live as part of a large global village with a variety of languages and customs, while still trying to keep everyone satisfied.

In general, we achieved this harmony by regularly meeting and coordinating with those in charge of discipline in each organization. Once a week, I met with all the sub-unit sergeant-majors to discuss any events that had occurred during the week, either operational outside the camp, or administrative inside the camp. In short, I ensured that all the company sergeant-majors (CSMs) and disciplinary staff were on net and headed in the same direction. Overall, we achieved our goal and I was extremely satisfied with the results. Our success in ensuring that close to 2,000 personnel got along while stationed at the same camp and sharing everything for more than six months is an achievement in and of itself.

Nonetheless, I faced several leadership challenges. First of all, at the national level, my CO was accountable to the Canadian commander in-theatre (the Commander of the National Command Element), from whom he derived all authority for specific operations. However, he was also answerable to the commander of the Kabul Multinational Brigade, who directed him to conduct certain tasks and operations. This situation sometimes put my CO in a difficult predicament as he was responsible to two chains of command. Granted, this situation is not unique. Even so, a great deal of personal diplomacy was required to accommodate both chains of command while remaining as operational as possible.

Another significant challenge lay in the need to work as a team with other contingents whose training and rules of engagement (RoE) differed from ours. It often happened that one contingent would be limited by its national chain of command in the execution of certain tasks while other contingents were granted greater leeway. Joint operations required an extremely high level of coordination.

Joint operations with the local Afghan police and with representatives of the Afghan National Army (ANA) posed another set of challenges. Proper execution on the ground required night and day rehearsals and coordination meetings. However, in an effort to maximize surprise and, especially, to ensure the safety of our task force personnel involved, it was also strongly recommended that local authorities and personnel participating in the operation be briefed at the last possible moment.

The Afghanistan tour of duty was our first experience with journalists embedded with the troops in the camp. In general, this cooperation did not cause any major problems. Journalists, of course, for security reasons, were excluded from operational preparations. They were also barred from participating in certain operations or missions, both for their safety and ours. Moreover, a meeting with the Public Affairs Officer (PAO) needed to be conducted before any interviews were granted. During interviews, troops were permitted to speak frankly while remaining within their field of expertise.

I truly believe that our mission was a great success. The highest military and civilian authorities in Afghanistan concurred with this assessment on several occasions. We made a difference in this country that has been devastated by decades of war. Our attitude and irreproachable behaviour convinced the local population and the ATA that we were a task force that constantly strove to improve the daily lives of Afghans, rather than simply another occupying force. Moreover, our success in bringing all our personnel safely back home after such a hazardous mission should not be belittled.

I sincerely believe that there is very little that our unit should have done differently, either before or during the deployment. Before receiving the warning order to deploy to Afghanistan, we had plenty of time for intensive and rigorous training, 18 months to be exact, since being tasked as the Immediate Reaction Force Land (IRF(L)). Furthermore, our training was systematic and relevant. Our CO had correctly assessed the importance of training in complex terrain, (specifically in built-up areas), which served us well in theatre. We had also received excellent briefings on various subjects, including operations, Afghanistan and the local population's customs and traditions.

I would advise my subordinates to continue to have faith in the chain of command. When problems or complications arise, deal with them quickly through the chain of command so that corrective measures can be put in place. These days, soldiers tend to try to solve serious problems by directly contacting agencies set up within the system instead of going through the chain of command. This behaviour, however, often results in frustration as the process can be stalled because the agency needs to send the file back to the chain of command since basic procedures were not initially followed.

To my peers, I suggest that patience is the watchword. Personnel administration is increasingly demanding and time-consuming. The system is very cumbersome to navigate and the numerous support agencies can be somewhat confusing. We must also ensure that we take excellent care of our personnel while remaining honest and discerning. The fact is that more and more soldiers, and even some senior non-commissioned officers (NCOs), are misusing and abusing the system. Strict, rigorous and sustained supervision will ensure that the unit functions smoothly and that our personnel are looked after properly.

To my superiors, I ask that they continue to trust the senior NCO corps, which is still the backbone of the Canadian Army. Unit warrant officers and sergeants spend most of their careers on the ground, be it in garrison or on exercises and missions. They are experienced and have seen it all. Give senior NCOs the chance to do their jobs and they will perform very well.

As for the Canadian Army as a whole, I sincerely feel that it has been stretched to the limit since the early 1990s due to our involvement in numerous and repeated missions. In this regard, I believe we must do one of two things: increase personnel or reduce our international commitments. On average, senior NCOs in today's Army have served on four international missions. In Afghanistan, I knew of two master-corporals who were on their seventh six-month tour and they had families waiting diligently for them at home. Within the past seven years, 3 R22^{e}R has been on four missions: Haiti in 1997 (at 70 per cent of unit strength), East Timor in 1999-2000 (at 50 per cent of unit strength), Bosnia-Herzegovina in 2001-2002, and Afghanistan in 2004. It is worth noting that when we received our IRF (L) warning order, which implied future deployment, we were already on a mission in Bosnia-Herzegovina with two more months to go in theatre. You can imagine that some soldiers and their families found it extremely difficult to face the fact that they would probably be redeployed once again in the near future. At this breakneck pace, in a few years we will have senior NCOs with six, seven or eight missions under their belts and this will be the new norm. Sadly, I do not believe that the families of these soldiers will be able to endure such extended and frequent absences.

One idea that seems quite popular is to shorten the tour of duty from six months to four months, as some national armies have already done (e.g. France, the Netherlands and Norway). But then again, that does not solve the problem of too few members. Tours will simply come up sooner unless we increase the number of soldiers in the Canadian Army. Furthermore, when I talk of increasing the number of troops, I mean the formation of new army units, not an increase in strength of existing units.

Currently, when a field unit deploys or is building up for deployment, it must pillage sister units to meet its deployment strength. That is what 3 R22eR had to do during the last two rotations. For Bosnia, we needed 108 personnel (all ranks) from 1 R22eR to reach operational strength. For Op Athena, approximately 100 troops had to be transferred to us. This shuffling of personnel causes a great deal of upheaval and seriously affects unit cohesiveness. Furthermore, the high frequency of deployments has resulted in an increasing number of personnel who are non-deployable for various reasons. It is currently estimated that between 20 and 25 per cent of our troops, in all trades, cannot be deployed overseas, a situation that I find very worrisome.

CHAPTER 11

COMBAT TOUR IN AFGHANISTAN: LESSONS LEARNED FROM AN RSM

Randy Northrup

In July 2004, I joined the First Battalion of the Princess Patricia's Canadian Light Infantry (PPCLI) as the Regimental Sergeant-Major (RSM). At the time, I was aware that the Battalion was preparing to deploy to Afghanistan. I had already been on several deployments prior to 2004, including Cyprus and Croatia, and I eagerly anticipated the deployment with the Battle Group. Of course, in my opinion, the pinnacle of any non-commissioned officer's (NCO) military career would be to serve as the RSM with a Battle Group in actual battle. And, indeed, the mission in Afghanistan was totally different than any previous mission I had been on. Nonetheless, my previous missions contributed to my sense of preparedness for Afghanistan. First, having been through the ranks and having previously deployed provided me with a certain degree of credibility with the troops. Young soldiers look to senior NCOs as the "old dogs" in the Battalion and ask "what did you do when you were a young fella?" You need to be able to answer, "when I was a young fella this is how I dealt with that" and based on the specifics of his concerns you should be able to give him some advice. You have to be ready to give advice on everything, whether they ask for it or not. In fact, a lot of times it is unsolicited advice. Having been on previous deployments helped me predict what types of things the soldiers needed know.

One of the most important things that I have learned from my previous deployments is the need to prepare troops to be mentally tough. When that hard time comes, you need to be ready for it. You cannot sit and wallow in self-pity. You have to get on with it. It is not because you are callous, rather there is a time and a place to reflect on hard times and, until then, you need to push on with your job.

This advice is particularly pertinent for RSMs. My tasks and responsibilities before, during and after the deployment with the Battle Group were diverse and demanding. There are a lot of responsibilities associated with being the RSM, particularly when associated with a Battle Group in a high risk area. My responsibilities ranged from professionally developing the NCOs below me – as they are the future of the army – to leadership training and proper management of soldiers in the various positions throughout the Battalion. Moreover, as the RSM, I was responsible for coordinating and working with the Operation's Officer of the Battalion on all the day-to-day issues. These included making sure that we identified and retained the right personnel for the Battle Group and that we applied the theoretical lessons learned in Canada to the theatre of operations.

Everything concerning the welfare of the unit and each individual soldier within the unit, particularly the morale of the unit, is of vital importance and one of the principle responsibilities of the RSM. For instance, sometimes small, simple changes can make a huge difference to the morale of a unit. Something as minor as changing the reporting time of morning parade from 0730 to 0800 can actually make a significant difference to morale. The later time allows children to be dropped off at school and traffic concerns to be minimized. Acknowledging that at times there is flexibility to the schedule, and that individual well-being helps to guide events, allows people to connect with and appreciate the group more.

Certainly, the responsibilities of an RSM run the whole gambit. You worry about the soldiers' personal well-being, their financial well-being, whether they get sufficient kit and equipment and whether the leadership that they are currently serving under is doing what they are supposed to be doing. Moreover, as RSM, I can speak to the Commanding Officer (CO) on all facets of the Battalion and all aspects of training. I talk to him about his officers; I talk to him about the other NCOs; it is just a smattering of everything.

In the end, I answer to one person. I have hundreds of lines of communication with every officer below the rank of lieutenant-colonel and senior NCOs and non-commissioned members (NCM) but I answer only to the lieutenant-colonel, the CO. My advice to him is either accepted or denied based on my technical and tactical experience.

Trust and honesty in the CO – RSM relationship are of vital importance. As the most senior NCO of the Battalion, I think it is important that the CO and I have a relationship that is honest and trustworthy. If you can establish this type of relationship with your CO, then there is nothing that you cannot do as a team. Nothing! The Battalion will certainly benefit from that.

Thus, in order to effectively transition from a senior NCO to the RSM, you have to be true to yourself. In my mind, it takes me longer to shave in the morning than it does to motivate my staff to come into work. I believe that soldiering can still be fun even in very arduous times – and it should be.

My advice to NCOs is to first off be honest and loyal, and not necessarily in that order. You need to be honest with yourself and everyone else. You also need to be loyal and to instill loyalty in the chain of command. We say "mission, buddy and then self." It is mission first, you look after your buddy second and you look after yourself last. Never will your problems be greater than those of the people that are below you in rank and/or position. The soldiers need to know that their superiors are there for them 100 per cent of the time. That is the job of an RSM.

I was very fortunate to have been the RSM for 1 PPCLI for two years prior to going into theatre. Consequently, I knew that the members of the Battalion, particularly the soldiers, had received the utmost attention regarding training and I knew that the officers were being closely monitored and guided by the CO, Lieutenant-Colonel Ian Hope. So I knew that, for the most part, I did not have to worry about soldiers and officers being unable to do what was expected of them. The apprehensive part of our pre-deployment training was not the 80 per cent we knew; it was dealing with the other 20 per cent – the "unknown."

Within a few months of being deployed to Afghanistan, we had a taste of the unknown and it was not pretty. In order to combat this element, I had promised my CO the best NCO Corps we could build given the time restraints. We did that. Lieutenant-Colonel Hope was happy with what we delivered and I would have to say that a lot of our successes were based on our training and the expertise and abilities of the NCOs as well as the officers. Mental toughness was what got us through the tour.

Some of the unknown included receiving new equipment just prior to our deployment and thus arriving in theatre not quite sure as to how it would work in that environment. Canada is our training ground and we wondered how differences in climate would affect us and our equipment.

For example, if the Lord had taken every rock, every unwanted rock in the world, and put them in one place, then he put them in Afghanistan. The ground was treacherous to say the least. It was absolutely hard and our vehicles and equipment have never been put to a test like that before. We wondered if our vehicles were the best for this new type of terrain. As it turned out, they worked quite well.

Another aspect of our mission that was unknown was how we would be received by the local people. First off, we walked into a country where the locals generally do not differentiate what we wear on our uniforms – the Canadian flag – from that of any other of the coalition countries. Many Afghans did not even know what their own flag looked like so it was hard to expect them to recognize national differences among coalition members. To them, we were just another force coming through. I had a sign on my desk which said: in Afghanistan it took 35 years to create a warrior; it will take a lifetime to create a citizen. I believe that sign to be accurate. The question is thus, how do we, as Canadians, fit into this picture? We did not know how we would be received or if we could backup what we promised the Afghans.

To follow through on your word is the Afghan way. That is their philosophy, what they believe. If you say you are going to do something in Afghanistan, then you had better follow through or you lose all credibility.

Thankfully we had received some pre-deployment environmental training regarding the beliefs of the Afghan people, including their religious beliefs and their tribal structure. You can hear all about it in briefs beforehand, but it does not necessarily make sense until you are living it. We wondered how we would actually react in the face of the enemy. We questioned if our training had adequately prepared us to do the right thing at the right time.

Once we actually got involved in combat operations we knew we had gotten it right. That 20 per cent unknown diminished to about a 5 per cent factor. Nonetheless, there remained a little bit of unknown: where is the next improvised explosive device (IED) going to strike; where is the next ambush; and when we go back into this town to check on the locals, is that school still going to be standing and are those kids still going to be alive. Under that first heat of battle, those are the types of questions that still remain.

When we first arrived in theatre we quickly started to go from our staging area at Kandahar Airfield out into our various areas of operation. Consequently, the local Afghans were very apprehensive, very untrusting of our movements. It is important to remember, however, that they have seen foreign forces enter their country time and time, and time again. Nonetheless, Afghanistan is a country that has never been dominated by a foreign force. Somehow the invaders always leave. To them, the British were defeated twice and the Russians were recently beat out and now there is this new monster coming out of the dust, dressed in helmet and goggles and all the other fancy equipment.

We were in their backyard. There was no electricity, no telephones, no radios. Everything was done by word of mouth – that is their method of communication. And, yet, we were trying to tell them about their country, trying to encourage them to support the new Afghan government. We were telling them what their government was trying to do for them. They had no idea who the president was or who the governors were. Their world evolves around their village elder. To us, it was like walking back into the Flintstone's era. We would drive down a road only to find a small community living in a cave carved into the side of the rock face.

Our lives were so foreign to one another. At first, all the Afghans hid from us. Slowly, the elders would come out to ask for certain things. We gained credibility through our local interpreters. Then, gradually, the Afghans started to recognize our faces and to gain some trust in us as we made good of our promises. It was only through that physical connection, for example, shaking their hands and sitting down with them at a Shura, that trust was built. Only then would they provide us with information like, "oh by the way we had six guys in the village last night and they took this, this and this and two young men and they are supposed to be back in two nights." That kind of information would help us plan our operations.

Eventually, the Afghans warmed up to us. In fact, it was not uncommon to see soldiers kicking a soccer ball around with Afghan kids while the local adults chatted. Unfortunately, however, if we did not keep our promises, then all the trust we had built up was instantly eroded.

The nature of our mission meant that fear was omnipresent. We were unable to succumb to it, however. One of the best ways to overcome fear is through mental toughness. This skill can be built up during training. The soldiers were briefed on the mission and told what to expect. One of the things that they were told was that it was very, very possible that not everybody would be coming home. Of course, a statement like this would make the bravest of men somewhat apprehensive. For instance, I know I am not bullet proof. It is how you deal with your fear that matters, however.

For the soldiers of the Battle Group, the more exposure they had to intense, stressful situations, the easier it was for them to deal with fear and all the uncertainties associated with the situation. The more they dealt with the unknown, the more they knew what to do, how to react and could predict what was coming next. After a while we got in-tuned with the environment and we learned how to recognize the precursor of a firefight. Paradoxically, continual exposure to these types of stressful situations helped to keep fear at bay.

We became hardened to the stress. The way we lived contributed to that too. We lived like rats. We had three tents that could hold up to 200 personnel each. Each tent had between 168 and 190 people living in them. There was no escape. You needed to depend on your fellow soldiers for everything.

Soldiers knew that this relationship held true in the field as well. Whether they were a casualty or witness to an incident that had casualties, because of the strong leadership they knew somebody was looking after them. This knowledge helped to minimize the fear factor. Knowing that that Lieutenant-Colonel Hope would stop operations even if we were in a firefight to extract all our wounded personnel before carrying on really contributed to fostering good morale. All the soldiers knew that they were important and that they would be taken care of no matter what the circumstance. That assurance was a great force multiplier. Those are the kinds of things that chip away at the fear factor.

Unfortunately, Afghanistan will never be independent until Afghans are able to provide for their own security through institutions like the Afghan National Army (ANA) and the Afghan National Police (ANP). They are trying to build these institutions as we are trying to build trust and to work with them. A lot of the people that we worked with who were part of the Afghan National Security Forces were Mujajadeen. They had fought the Russians. Notably, they had been paid well by the United States to do so. They also knew the potential of the country.

Nonetheless, working with the Afghan National Security Forces was difficult because they lacked the necessary equipment. For instance, just to keep track of them we had to give them glow sticks at night so that our people could see them. We have night vision; they do not. We have uniforms; most of the Afghan Battle Forces, as we called them, had no uniforms.

We would have loved the opportunity to work with the ANA, however, we were not so fortunate. Conversely, the ANP have a reputation for corruption and they are not obligated to stay in any one area. Moreover, police action and army action are two different things and their roles cannot be collapsed.

Yet, in some areas, only ANP were available and, under the circumstances, we were forced to work with them. For example, during our rotation, there were 1,200 ANP in Kandahar Province, with approximately 700 ANP personnel in Kandahar City alone. Not all of them were registered police, however. Some of them were a sort of reserve force. Yet, that is who we had to work with. As far as uniforms were concerned, some ANP had a jacket; some had pants; none had both. Adding to our difficulties was the fact that we were army personnel trying to teach police how to be police. That should not be our job. Our job should be to train the Afghan military so that they understand what army action is all about. With no ANA presence, however, we resorted to training the ANP.

As mentioned though, working with the ANP was frustrating, particularly because they would not hold ground. Each time we took the same piece of ground it would cost us and, I assure you, it is a very expensive price when you lose a soldier taking the same piece of ground over and over and over.

Issues like the aforementioned greatly frustrated our efforts at providing humanitarian aid and reconstruction. You cannot provide reconstruction – construction, as was more often the case – without security. With limited resources, concurrently providing development, diplomacy and defence was often a challenge. Moreover, if you said something against a local governor, he would take it personally and disagreements on tactical points eroded the working relationships we had fought hard to establish.

Lieutenant-Colonel Hope took a hard-handed approach to these issues. He assured the Governor in our area of operation that he and he alone commanded his forces. Also, for security reasons, he would not provide the construction teams until fortifications were established.

The Afghans understood this type of forcefulness. Soon Lieutenant-Colonel Hope and the Governor had a relationship built on mutual respect. In return, we gave him the joint coordination centre and we provided him with people to assist in making him a better governor.

Unfortunately, in terms of development, everything we did became a prime target for the Taliban to destroy and thereby undermine our role in Afghanistan. Once you promise something to an Afghan, if you do not keep your promise then you lose all credibility in their eyes. The lack of security in the area greatly challenged us on this point. Even when we imposed security in the area, it was next to impossible to maintain it. In the end, what is needed is to put an Afghan face on security and allow foreign troops to help with (re)construction.

In the meantime, it is important for our forces to know exactly what they are shooting at. Do not just fire over the fence, you do nothing but empty your magazine. Aim and then fire. That is what we practiced during training. Once we got through the first couple of ambushes, we realized we stood a better chance of winning by fighting our way. We became battle-hardened. No one is going to shoot real bullets at you in Canada and no one is going to die or be dismembered on your training event just to see how you react. In the field, however, things are totally different. You need to react appropriately when a rocket propelled grenade (RPG) round hits and both the legs are ripped off the soldier next to you or a buddy catches a round right square in his chest. You cannot train soldiers for that. You can only make them aware of it and prepare them as best you can for such an eventuality.

In this respect, listening to our veterans of the war in Afghanistan is invaluable. Moreover, it also helps the veterans deal with what they have gone through by talking with sympathetic comrades. These young soldiers who are returning from Afghanistan are the new Canadian warriors. The more we talk about it, both casually and through professional development sessions, the more we can learn from the experiences of these young men and women. It would be a crime not to garnish all the knowledge and experience that we have learned through the last several rotations and pass it on. I tell the NCOs and officers in the Battalion that anybody that was over there can put up his hand and say "I did my duty." If you are still in uniform, however, you are not done your duties. You have more responsibilities, one of which is to train the young men and women who are going over next. We also have the duty to remind Canadians of the sacrifices that we make for them. It is not the Canadian soldier's job to tell you why we are there; rather, it is our job to tell you what we are doing there.

So, to all those veterans, I must repeat, when you return from rotation, your job is not finished. You still have duties and responsibilities here in Canada. Part of that is making sure the next batch of soldiers going over get every tidbit of information that they are willing to listen to before they go.

Luckily, the purpose of the mission itself helped to build the morale of the troops. There are a lot of soldiers in the Battalion who realize that there is still more to do in Afghanistan. They feel that their jobs are not over yet and they want to return on a future deployment. It says a lot about the mission when men and women who have witnessed what our veterans have in Afghanistan want to return. For many, this is exactly why they joined and they feel like they can make a real difference in the world. Veterans who return can provide a group of rookies with that pillar of knowledge and experience to help finish the job.

To conclude, when I was appointed as RSM of the First Battalion, PPCLI, I did not think that it could get any better. This position is the pinnacle of my rank. Even more important, as an infanteer, I thought it was the pinnacle of my career to be an RSM with an infantry Battalion. I was mistaken.

When I went overseas I realized that the highlight of my military career was being the RSM of a Battle Group in battle. I will always cherish that experience. I'll always remember and honor those people that were with us. We can never forget our veterans. We must always remember where we have come from and those we have lost along the way.

CHAPTER 12

WORDS OF WISDOM FROM A TASK FORCE RSM: OPERATION ATHENA, ROTATION 3 – FEBRUARY TO AUGUST 2005

Lee Topp

Operation (Op) Athena, Rotation (Roto) 3 was mounted with troops mostly out of 2 Canadian Mechanized Brigade Group (2 CMBG) and Land Force Central Area (LFCA). Notably, the Task Force Commander, Colonel (Col) Walter Semianiw and I, the task force Regimental Sergeant-Major (RSM), were not from the parent organizations.

As the RSM of the Armour School, I had just arrived back in garrison after two weeks in the field with an Armour DP 4 (Squadron Sergeant Major) course. The commanding officer (CO) of the School informed me that they were looking for an RSM for the Task Force in Afghanistan and asked if I would like my name to be submitted. I said yes. In fact, I had been asking for a mission for some time. The next day, which was the Friday of the Thanksgiving weekend, I was informed that I was selected to go to Afghanistan. The reconnaissance (recce) party was to leave on Thanksgiving Monday from Trenton. I quickly determined that time, space and activities would not allow me to meet the recce timings and be in Trenton by the flight date.

I was given the Commander's name and phone number. Subsequently, I phoned Col Semianiw and informed him that I would be unable to attend the recce. He informed me that the Commander of 2 CMBG was heading to Afghanistan for a visit the following week and suggested that I join his party. I made some calls and was able to join the trip the following week. Although, this would not allow me to meet Col Semianiw, or any of the other staff or command teams on the recce, it would give me the chance to see the theatre. I knew that, unfortunately, not being on the recce with the main group would mean that I would be behind the eight ball when it came to the fostering of common shared experiences and setting the tone of the mission right from the start.

As I hung up the phone, I realized that this would be the second mission that I would execute on short notice and with a group of unknowns. A decade before, I had headed out the door on two weeks notice as the RSM of a unit that had not even been formed. At that time, I had had only a few days with the CO and almost no time to meet any of the unit before heading in on the recce, from which I did not return until the end of my tour. The unit joined us three weeks later. Having spent all of my time up until that point on Regimental Duty learning about personalities and capabilities of those that I thought I was being deployed with, I was surprised when I headed out the door with a group of strangers. This seemed like an odd way to do business then and it still strikes me as peculiar today. Nonetheless, I was glad to have the chance to head out on an operation again.

Immediately, my mind turned to the "should haves," "could haves," and "would haves" that I had learned from my last tour. I could already feel apprehension creeping into my life. I simply pushed the thoughts away. Accepting the challenge, I took a deep breath and carried on. I started to make a mental list of things that would be needed, beginning with requirements for my first day in Petawawa, and I started collecting as much information as possible about my upcoming mission.

I gave a great deal of thought to how to set the tone and provide the direction and guidance that would be expected of me. I consider myself to be direct in approach and professionally competent. I have the same expectations of others derived from having spent time either in a tank troop of an Armour Battle Group or, for the most part, in recce at the Brigade and Regimental levels. I am comfortable working on my own or within small groups on the outer fringe of the area of operation (AO), making my own decisions. Moreover, I am willing to accept the risks and responsibilities associated with the decisions that I make.

The first time that I met Col Semianiw was in the airport in Kabul, he was on his way out of the country after the recce when I arrived. We had about 20 minutes to speak and he directed me to take care of the RSM issues. Then he told me that he would see me in Petawawa no later than 1 November 2004. This short meeting between the Commander and I was insufficient time in which to form a personal opinion of him. Not that this was yet important. Admittedly, however, I felt like an outsider to the recce team because I had not been with them.

THE RECCE

Op Athena, Roto 2's main effort had been providing support to the International Security Assistance Force (ISAF) by providing the Recce Squadron to Kabul Multinational Brigade (KMNB). During their tour, Afghanistan had conducted Presidential Elections. Roto 3 would have the same main effort during the National Assembly elections.

Op Athena's Roto 2 Task Force RSM, Chief Warrant Officer (CWO) Chris Kitching, was a long time friend of mine and I had been in contact with him since before my nomination. I was not concerned about the details of the operation, as much as getting a feel for the mission and the country. The rest of the information I could get from the Deputy Chief of the Defence Staff Directives for International Operations (DDIO), Task Force Standing Orders, operational plans and any of the other pertinent documents.

At the beginning of my tour, I spent a day with each of the principal units to get a feel for their organization, looking at their tasks and listening to recommendations for change. It appeared right from the start that this was an eclectic organization.

The Recce Squadron (Sqn) had been augmented with an Infantry Recce platoon, sniper detachments and a platoon of Slovenian troops. The Sqn had an operational strength of about 200 personnel, including the Slovenians. The Slovenian troops were well trained, having special mountain operations training and were well respected by all in the Task Force.

Recce Sqn was the only declared asset to KMNB. It operated throughout the entire KMNB area of operation. This area was approximately 5000 square kilometres of diverse terrain, which included, the city of Kabul, open rolling plains in the north and extensive mountain terrain in the south, east and west.

The Infantry Company had three platoons and was augmented with a platoon of Belgian troops. Together they provided the security of the camp, quick reaction force (QRF) duties and they patrolled the two-kilometre bubble around the camp on a daily basis. Additionally, under an agreement with the Norwegian Battle Group which was responsible for that AO, the company occasionally patrolled the area south of the camp known as Cahar-Asyab.

The Engineer Sqn contained improvised explosive device (IED) and explosive ordnance disposal (EOD) capabilities, as well as a field troop and some heavy equipment capabilities. Although the Engineer Sqn was not a declared asset, it did a rotation within the KMNB engineer units to provide a first response capability for IED/EOD incidents within the Kabul city area. This aided KMNB when the other engineer assets were deployed with the battle groups. They also provided engineer support to Recce Sqn for route designations and basic engineer support.

The National Support Element (NSE) provided the normal maintenance, supply and transport functions. It oversaw the operation of the camp and civilian contractors. With a strength of over 200 personnel, it was one of the larger units in the Task Force (TF). Grouped under the NSE, the civilian component from Canadian Forces Contractor Augmentation Program (CANCAP) and Canadian Forces Personnel Support Agency (CFPSA), numbering almost 400 personnel, provided camp support operations.

The National Command Element (NCE) held all of the staff and command functions, including: the Canadian staff and liaison officers at all coalition headquarters; the All Source Intelligence Centre (ASIC); the Health Support Service which provided a role 2+ hospital; the Military Police; NIS sections; and civil-military cooperation (CIMIC) and Embedded Training Team (ETT) functions. The ETT, later to be renamed the National Training Centre Detachment, was a 25 personnel detachment working with the US Forces in training the Afghan National Army.

The units making up the TF provided the Commander with Canadian assets capable of dealing with all, or at least most, tasks and situations that would be encountered. Thus, the TF was not dependent on foreign support.

During Roto 3, the Theatre Support Element (TSE) in the United Arab Emirates (UAE), which provided the air link to the mission, was placed under command of Task Force Kabul (TFK). The TSE CO was downgraded to a lieutenant-colonel and the TFK Commander became the Theatre Commander.

The third night of the recce, CWO Kitching and I were sitting in the mess discussing some issues when everyone except us jumped up and ran out of the mess. I looked at Chris and asked, "was it something that I said?" Rather, it turned out that the camp had come under rocket attack and we, perhaps because of our hearing impairments, had heard neither the attack nor the alarm.

This occasion, combined with a series of other incidents that required the deployment of the Quick Reaction Force (QRF) and other TF assets, allowed me to witness the procedure that the Task Force used to deal with these issues. It occurred to me that what we were dealing with was a scenario-based operation. The critical incidents could be broken into a number of areas or tasks and these could be used as a basis for training, and testing our Task Force, particularly, if used in conjunction with the standing orders and operational plans that were already in place.

During my time in Afghanistan, the strategic recce for the Provincial Reconstruction Team (PRT) was in theatre. They were travelling around Afghanistan looking for a location for Canada to mount a PRT. On my departure from theatre, the aircraft I was on flew to Herat to pick up the team and then took them to Kandahar. This was the first indication that there were changes coming in the mission.

Without the information or direction that had been given or gleaned from the Commander's recce, there was little more that could be accomplished. Once back in Petawawa we would have the chance to talk and compare notes. I returned to Gagetown, packed my kit and headed to Petawawa, arriving at my final destination by the end of October.

MOUNTING THE MISSION

Once the Commander and I had had a chance to speak with each other, it became clear that this upcoming mission and its command structure would be different from what was considered "normal" at that time. The Commander and I were, respectively, the NCE Task Force and Theatre Commander and RSM.

In Petawawa, the Commander and I had a chance to sit down and discuss issues and I was consequently able to gain insight into his impression of the mission and what he saw as the centre of gravity and the way ahead. As the two of us were unaccompanied, we finally had some time to get to know one another. For me, it was important to understand his life experiences because it would help me to better understand his thought process and what shaped his decision-making. My intent was not to become a "yes man." Rather, I would provide an additional set of eyes and look at each situation from a different angle of attack therefore hopefully providing us with more options.

There were issues that came from his recce that needed to be addressed. The most pressing was that the infantry company was to be reduced to two platoons and the Belgian contingent would be leaving the mission by the time we arrived in theatre. This would leave insufficient personnel for the force protection tasks. This deficiency was addressed to both Land and Joint Staffs. At this point, there was some discussion about how to mitigate this shortage from within the TF. The removal of some of the Infantry Recce from Recce Sqn was discussed, but I personally did not support this idea because Recce Sqn was the only declared asset. My thought was to look at reorganizing some of the other units in order to free up positions and to leave the recce asset intact. With assurances that we would eventually get a third platoon, the direction provided was to start training with three platoons. This was the only major change to the manning; the other suggested positions were minor in nature and impact.

I was briefed about the PRT because there was a chance that the PRT would be mounted sometime during our tour. At the same time, I was informed that the mission might change location either at the end of our tour or perhaps during the next rotation.

The NCE (at least the part that was from 2 CMBG) stood up two weeks ahead of the Task Force. The NCE took command of the Task Force on 12 November 2004. It shepherded the TF through its training and deployment. To my knowledge, this was the first time that this had been tried and it did present some problems. On the plus side, this arrangement provided the NCE with a chance to work together prior to deployment and to work out some of their procedures. It quickly became apparent that the command authority for the NCE would originate from the Deputy Chief of Defence Staff (DCDS) upon deployment. Thus, the Commander would have no real authority until the deployment and transfer of command authority (TOCA) had taken place. This was not a showstopper, however.

One of the first requirements was to meet the other command teams face to face so that we could get to know one another. Once the Commander gave his intent, it allowed the staff to commence work for the TF. There would be little time to become familiar with the command teams and staff prior to deployment. It was hoped that by the time we headed into the mission area that we would be at least comfortable with those command teams. Nonetheless, this was not the ideal way to head out the door.

The Task Force stood up on 12 November 2004 with a parade, followed by a town hall meeting. At the town hall meeting the command teams from each organization were introduced. The Commander spent some time in a detailed briefing of the mission and in describing the road to Kabul. Approximately 80 per cent of the Task Force was at this meeting. (Some specialized trades were still conducting training and would join us later in the training or once deployed.) Lieutenant-Colonel Fortin, the Roto 3 CO of the TSE, was also present and we had the chance to meet him for the first time. The TSE would complete their rotation by mid-December in order to be settled by the time our rotation would take place in late January/early February.

There was little doubt that those who attended the initial briefing were motivated and interested in the mission. This was the same Brigade that had lost three members on Roto 0. This loss had established a sense of seriousness about the training and the mission. At no time was it necessary to state the need for taking things seriously. If anything, there was little humour or levity. This extreme was not a good sign either. In fact, there were some that appeared to be uneasy to the point of showing fear and apprehension about the mission. This anxiety was something that training would hopefully fix. We would need to keep abreast of this issue as we headed down the road to Kabul.

The training conducted by 2 CMBG included theatre mission specific training (TMST), Peace Support Training Centre classes as well as a validation exercise. During this time, a number of family briefs were conducted. One issue that became apparent was the need for a French brief. There were more than 40 personnel who were French and even more family members who preferred to have the briefings in French. The rear party was well established and spoke at each family brief so all could put a face to those that they might need to deal with in the not so distant future. The Commander and I also attended and spoke at every briefing so the families could recognize us. The briefings contained pictures that were taken during the recent recce and illustrated the living conditions, work areas and the city of Kabul. Those from outside 2 CMBG were given a CD-ROM with all of the information that had been presented during the briefings so that they could take it home to their families. Great pains were taken to ensure that all the details of dealing with the families were taken care of before our departure. Perhaps as "augmentees" to the Task Force, the Commander and I could more closely identify with these issues.

The validation exercise took place in the Petawawa training area the last week of November and the first week of December 2004. This was the first time that the TF was in the same location at the same time. It was also the first opportunity to get to talk to the unit sergeant-majors. I took this opportunity to have a formal orders group (O group) so we could talk through some issues and I could set the tone for the way ahead. The location for the O group was in the NSE lines which had been offered by CWO Traclet, the NSE RSM. When I arrived, all were seated around a table waiting for me. The funny thing was that there was no chair for me. I quickly looked around and decided that there was one of two ways to deal with this situation. I could either raise a stink about not having a chair or I could remain standing and give my points from the floor. I chose the latter option.

As I began, the group began to act as though things were out of order. This uneasiness was understandable. Since my absence on the recce, CWO Traclet had become the defacto TF RSM. I obviously had work to do.

My direction to the unit sergeant-majors was simple. I like, most NCOs, live in a black and white world. The Code of Service Discipline would deal with all breaches of the Code. My message was to follow the rules, policies and procedures unless they made no sense and then to be prepared to be held responsible for those rules you break. My direction to them was to state their issues and to be clear about what they were going to do about it. I made it clear that I was not there to solve all the problems. It is not that I am afraid to make a decision, but if others want to influence that decision, then they must have a solution or accept mine without question. I also told them to ensure that those directly below them did their job and to insist that those below them in turn do the same. My message was clear: "Look at your troops and tasks with a critical eye and ask yourself why are we doing this? Is it because it makes sense and works or is it because we have always done it this way? Listen to yourself when you talk. If your conversations begin with I or me instead of us, we, the team, or the unit, then consider changing your language."

As we approached our deployment date, the weather was typical of Petawawa, cold with plenty of snow. In order to conduct training, a mock-up of the camp in Kabul was replicated. This was a huge undertaking for 3 RCR and took them weeks to setup. It was constructed of modular tents for the work areas and quarters, with a replica of the front gate in Kabul and mock bunkers. In my opinion the utility of the camp was questionable for several reasons. First, during the exercise, it was difficult to support the large infrastructure. Electrical and heating problems plagued the exercise, at times leaving parts of the camp cold and dark for days. In theatre, all of these tasks are taken care of by the civilian contactor so the TF is not configured to deal with them. Making the repairs and dealing with the problems took time and resources away from the main effort, the exercise and I could see little utility to this process. Second, this venue forced the events of the exercise to be played out inside the training area. This was of little help for a Task Force that would be deployed to Kabul, a city of two to three million people. Perhaps a location outside of the base, which could have better replicated the time and space of the theatre, would have been a better choice.

Moreover, shortages of some equipment meant that civilian vehicles were used to replicate actual in theatre vehicles. The shortages of these military vehicles prevented drivers from gaining much needed experience. It also meant that some of the drivers would not receive their driver training, especially on the "Bison," Light Armoured Vehicle Wheeled (LAVW), and Light Support Vehicle Wheeled (LSVW), until after the exercise, thus only days before deployment. The impact of this lack of training was apparent once deployed; our accident rates soared on initial deployment.

The Recce Sqn would have benefited the most from an exercise outside of the training area. It would be entering a huge AO in theatre, while working with the three different battle groups. In theatre, the Recce Sqn would be away from the TF camp for the majority of the tour of duty. However, the close proximity of the Recce Sqn during the exercise gave a false impression of when Task Force elements could or should be used to support the Sqn. This would require some adjustments to the TF operating procedures once in theatre.

All units were joined by some of the additional personnel during this training period and the Commander and I made a point of meeting each new arrival. The TF was at about 90 per cent manning at this time.

At the end of the exercise, most individuals felt comfortable heading into the mission, although there were some who still had concerns. It was not necessarily that they said so; rather a few had a steady look of uneasiness about them. Perhaps the exercise only highlighted their weaknesses and did not better prepare them. Only time would tell.

During the exercise, some of the staff were preparing for the deployment of the Task Force. This placed an additional burden on the staff as they used their down time during the exercise to complete operational tasks. As things progressed, some of the staff were cut from the exercise to ensure that the final details had been completed before the TF headed off on their leave immediately prior to deployment.

Once the exercise was completed, the TF was brought together again and all members were briefed on the deployment plan. This was done so all would know their departure dates before they headed off on leave. The leave encompassed the Christmas period and the first flights were scheduled to depart at the end of January.

The Task Force was back on the ground by the middle of January for final briefings and some last minute training. At this time, we were issued the C7A2 rifle. Some of soldiers were uncomfortable with the new weapons, especially since it had not been used during training.

The Joint Staff briefings brought us up to speed on the situation in theatre. The PRT was still in the discussion phase, and the Commander suggested that it might be advantageous to have a staff officer sent into theatre to deal with PRT issue from our end. We were told that it would not be necessary as we would not have any dealings with the PRT should it be mounted. We were also briefed that the changes to the manning had not been authorized yet and told not to worry.

THE DEPLOYMENT

The advance party deployed on 27 January 2005. They arrived in theatre two days later. It was at this time that the TF was informed that we would not be given the third platoon for the infantry company. This meant that personnel would have to be cut from the TF almost on the day that they were to deploy, meaning that the TF would have to be reconfigured on the fly.

The main flight, which included the Commander and myself, departed on 02 February 2005, arriving in Kabul three days later. This travel period allowed for a day at Camp Mirage which provided us with a chance to speak to the TSE command team and staff.

From the start, the deployment was plagued with weather problems. The only flights to land in Kabul were the advance flights. The Kabul airport did not have instrument landing capabilities and the weather would not allow for visual landings. The crash of a civilian airliner into the mountains east of the city while attempting to land in Kabul, and the death of all on board, confirmed the poor conditions.

Consequently, all flights were diverted to Bagram because of the continued snowstorms. This required the ground transport to leave Kabul three hours before the flight was scheduled to arrive. The TF had only enough ground transport (Bisons) to lift one Hercules load at any one time. With two flights arriving each day, and a six-hour return trip, the road parties were travelling 12 to 18 hours a day for the first two weeks of the rotation. The winter weather conditions and the unfamiliarity with driving in the city meant that when our drivers took over, as part of the normal relief in place, our TF suffered many accidents. Luckily, none were fatal.

The handover of the mission was completed on 8 February 2005. A simple office ceremony with the Canadian ambassador, Mr. Christopher Alexander, marked the change of command. The rotation would continue until 21 February when all of our troops would be on the ground.

Once the ceremony had taken place, a meeting was called to sort out the manning issues. There were no options at this point because it was too late to reorganize the TF. The removal of some of the Infantry Recce from Recce Sqn was the only choice.

We received order that two section of the Infantry Recce platoon, which had been attached to Recce Sqn, were to be chopped to form the Force Protection Company. A section from the D&S platoon from Camp Mirage was brought forward in order to provide enough troops to fill the tasks that the infantry company had taken over. This was not a popular reconfiguration amongst the troops, especially since their parent unit had guaranteed the Recce Platoon that this would not happen. This episode continued to be a source of discontent for the entire tour. It also forced Recce Sqn to reorganize and work with reduced manning. The discontent in the troops was, in my opinion, fueled by a failure of the NCOs to accept the need to reinforce Force Protection Company and then to motivate their solders to accomplish the task. It is an NCO's job to instill pride in the soldiers regardless of what job they are doing. If the job is worth being done, then it is worth being done right. There was no more important job than providing a secure base location for the TF.

MISSION AND CAMPAIGN PLAN

The International Security Assistance Force (ISAF) mission was: to Assist the Afghanistan Transitional Authority (ATA) in maintaining security within the ISAF AO so that the ATA, as well as the United Nations (UN), could operate in a secure environment in order to allow the ATA to build up the security structure in Afghanistan, as agreed to in the Military Technical Agreement. We were instructed that "Task Force Kabul Roto 3 will assist the ISAF and Afghanistan Security institutions in ensuring a secure environment within the ISAF AO with a view to facilitating Afghan national development."

CAMPAIGN PLAN – LINES OF OPERATIONS

We were given specific instructions about what we were supposed to do. Our job was to: provide support to Kabul Multi-National Brigade; provide support to the National Assembly Elections; establish a robust force protection plan for Camp Julien; foster strong relations/ties with the Canadian Embassy, Department of Foreign Affairs (DFA), and Canadian International Development Agency (CIDA); and, if required, to support to the Canadian Provincial Reconstruction Team.

TASK FORCE STRUCTURE AND LIVING CONDITIONS

The TF numbered 747 all ranks in Kabul and 238 in the TSE. The troops in Camp Mirage were Air Force, with the exception of the D&S platoon which was Army Reserve. Of the 747 in Kabul, there were 117 officers, and

530 NCOs. Of the officers, 64 were Regulars from outside LFCA and 53 were Reservists from 29 different units. In total, there were 203 personnel who were on their first deployment, 140 on waivers, 54 females, 5 married couples and two family groups. The mission was supported by almost 400 civilians working for CANCAP and CFPSA.

Camp Julien, which housed not only our troops, but also troops from Norway, Hungary, Turkey, Germany, Slovenia and the US, later to also include Romanians and Italians, held from 1500 to 2000 personnel at any one time. Camp Julien was without question the best Camp that the North Atlantic Treaty Organization (NATO) had in the Kabul area of operation.

The camp provided a level of service seldom seen before in a theatre of operation. All had access to phones, the internet, computer gaming, gym facilities and messes. The camp produced and purified its own water. It had its own sewage treatment facility. It was surrounded by a two-layer height of Hesko Bastion which shielded a 2.5-kilometre running track inside the confines of the camp. It also had its own laundry service. The only vehicles, other than TF or NATO vehicles, that were allowed inside the camp were the food trucks. Fuel was pumped from outside, under the Hesko wall, where it was tested and then hooked into the bladders. Garbage from inside the camp was collected by our personnel and then placed in a compound that was separated from the main camp before locals could pick it up.

The camp was 1.6 km long and 0.6 km wide. The size presented a big target and was difficult to secure. It was located in the southern end of Kabul, in an area referred to as Dahrlaman, so named after the Dahrlaman Palaces which border it. The camp operated under a blackout policy at night, with the exception of the two gates which were the only entry and exit points.

Camp Julien was the only NATO camp in the south of the city and was separated from the other camps by a mountain range that bisects Kabul. This meant that should we require support from one of the other camps it could take thirty minutes or more for them to reach us. Our isolated location caused problems with communications, especially once the troops had passed to the north of the mountain range. Communications required a radio relay on the top of the mountain that bisected the city. Most of the TF business with other NATO forces was conducted north of the mountains. The airfield and all other headquarters (HQs) and contingent camps were there.

The high level of services provided inside the camp was because there was no rest and relaxation (R&R) or other opportunity aside from operations to leave the camp for members of the TF. Those that did not have an operational reason to leave the camp, in most cases, did not leave. Over time, a conscious effort was made by all to ensure that, if possible, those "shut-ins" were allowed to accompany others who normally travelled outside of the camp. This helped to lessen the fear of the unknown within the TF.

The camp, because of its design, created a mentality among the personnel that is best described as "a Fort Apache on the Bronx." Camp Julien provided a complete separation, both visual and physical, from those that we were to provide security to and interact with. One might be able to justify the level of protection by saying that the level of threat required it. I was asked one day what my opinion was on the level of security and my response was that I would have taken down the Hesko and replaced it with wire so we could see out and they could see in. Of course, this was not the expected response and I received many strange looks. I did not believe that there was a threat of direct conventional attack against the camp because we simply had too much firepower. Moreover, the separation from the outside world created "boogie men" for those stuck inside. Perhaps though, I have a different view on accepting risk?

TASK FORCE UNITY

The development and maintenance of unity within the Task Force was a constant issue. Unlike a traditional Battle Group with homogenous unity intertwined to provide the full spectrum of capabilities, this was a horse of another colour. Although all the elements of a traditional Battle Group were present, units did not work together and even in some cases had little or no contact with other elements of the TF.

Surprisingly, an education programme was required so that units could identify and appreciate what it was that others were doing. The first issue occurred when some of Recce Sqn started calling the gate guards from Force Protection Company "commissionaires." This was not well received, particularly given the reorganization that had happened between the two units. This potential problem was nipped in the bud as quickly as possible and all units were told that berating other units would not be tolerated. I was convinced that this negativity was a sign of immaturity and poor leadership. Unfortunately, this incident was not the last time that I needed to step in and control such an issue. Indeed, each time it angered me more as I expected more from the troops.

On a different occasion, a decision was made that Recce Sqn would not get a fresh ration supplement when deployed on hard rations. This decision was made even though they would spend as much as ten days at a time out of the camp. The justification for such actions was that we had overspent our ration allowance – or something to that affect. Once again there was a need to put things into perspective. I suggested that if there were insufficient ration for Reece Sqn, then perhaps we should stop happy hour and coffee break food. If that did not free up enough ration, then perhaps the camp should go to two meals a day. As one might imagine, the problem went away.

There were some units, because of security issues or job type, that had almost no interaction with the TF. At times, even up to midway through the tour, I saw unfamiliar faces wandering around the Camp. This occurred despite the amount of effort that was put forward to ensure that the Commander and myself met all new arrivals. Coupled with a number of Staff Inspection Visits (SIV), Staff Assistance Visits (SAV), technical assistance visits (TAV), recces, visits and replacements, there was a constant flow of "unknown" personnel into and out of the theatre.

The Recce Sqn was undoubtedly the envy of all given their mission, freedom of movement and large AO. At times some forgot that we were in Afghanistan primarily to support their operation. Additionally, some resented the priority that they had on resources.

TASK FORCE TASKS

Upon arrival in theatre, the Commander had made a conscious decision to allow the units a two-week period to get themselves sorted out without interference. The units were tasked to conduct a Force Protection review of their unit operations. At the end of the two-week period, they were to provide a brief on any things that may have to be changed, modified or amended.

The staff spent this time reviewing and updating the operational plans. The two weeks also allowed time for the Commander to meet those personnel new to the TF for the first time as well as the civilian staff from CANCAP and CFPSA. Meanwhile, the NCE was busy getting into rhythm and coming to grips with the daily requirements of reports and plans.

This mission was an intelligence driven operation. The All Source Information Centre (ASIC), with all of its resources, provided an overwhelming amount of daily information. It had far-reaching national capabilities, and the ability to reach into other nations' resources. ASIC was considered one of the most capable organizations in theatre. It provided both the TF and the Embassy with insight into what was happening in theatre. As the mission continued, we would soon realize that a method of filtering the vast amounts of information would be required.

Recce Sqn was already in full swing with the new Turkish HQ's at KMNB. The Slovenian troops, because of their mountain operations training, had been tasked to look for survivors of the civilian aircraft that had recently crashed. They were airlifted, by Chinook helicopter, close to the top of the mountain where the crash had occurred. Dropped off some distance from the actual site, they had to march the rest of the way. The snow was estimated to be 14 feet deep at the summit. Sadly, they found no survivors. When the Afghans authorities requested the use our troops to conduct the body recovery, we declined. Later in the spring when the Afghan National Army attempted the body recovery, mines that had been placed on the mountain killed three soldiers.

Recce Sqn was involved in Operation Horseshoe, an operation designed to provide over watch throughout the capital region in order to prevent, or at least reduce, the number of rocket attacks into the city and NATO camps. Known launch sites as well as access and escape routes to those sites were monitored. Several of the sites were just south of the camp, looking into the Jowz Valley, which provided an additional level of over watch for the camp.

Force Protection Company was reviewing its tasks and getting settled into the job. They were conducting gates guard, mounted and dismounted patrols within the two kilometre bubble around the camp and practicing the QRF procedures. This company eventually had the greatest input in the force protection review.

The Engineer Sqn was completing the final stages of training to ensure that the IED teams were ready for their rotation as KNMB first responders. Snow removal became a task that none of us had contemplated before we had left home.

The NSE was occupied with the rotation as well as the daily operations of the camp. The cold weather that hampered the rotation caused difficulties with getting fuel as the mountain passes were closed because of snow. The daily flights of personnel and supplies kept transport (tpt) on their toes and supply and maintenance hopping. The NSE staff quickly became familiar with the civilian contractors and their operations.

Once the force protection review was complete, it provided 85 recommendations for changes, improvements and amendments. These were prioritized and then tasks were given to complete those that were achievable.

One of the things that I took issue with was the blackout conditions in the camp. Although not mentioned as one of the 85 points, there was, in my opinion, no added security provided by this tactic. The front and rear gates were clearly illuminated giving a left and right of arc for targeting purposes. Moreover, there had already been a number of personnel who had hurt themselves falling in the blackout conditions. The question was asked if we are hit with a rocket attack do we have emergency lighting to search for and then handle casualties. The answer was no. Finally, the camp, being located in the southern fringe of the city, was in an area which was slowly coming alive with lights at night, thus leaving the camp to appear as a black hole. The recommendation to forgo blackouts was not immediately popular. Once we turned the lights on, however, and no rockets immediately followed, everyone agreed that it made moving around the camp at night much easier and safer.

Battle Rhythm

With the TF rotation complete, it was time to establish a rhythm for the tour. Each morning would begin with a Commander's Conference, which provided an intelligence update and a chance to be briefed on TF activities for the last and next 24 hours. This was followed by a look at any long range issues and a round table discussion. Then there was a trip to the ASIC for a detailed intelligence brief.

Weekly, the NSE command team would sit with the Commander and me to discuss any issue that needed to be addressed. (The NSE Commander and RSM would replace us during our leave and were always kept in the loop.) During the week, the Commander would be briefed by medical, legal, religious advisors and civilian leaders from CANCAP and CFPSA.

On Saturdays, the Commander's Conference was expanded to include representatives from Canadian Staffs at the different HQ's in the city. Once a month, a large conference was conducted with a representative from every unit or HQ that had Canadians in theatre. The Commander and RSM of TSE would travel into theatre to attend these monthly conferences. Later in the tour, when the Theatre Activation Team (TAT) for the PRT was in theatre, the Commander and RSM of TSE also attended this meeting. It was a chance for all to see the issues that were being dealt with across the theatre.

Sundays did not bring much relief. At the sub-unit level, they attempted to manage some breaks but even this only worked for some units and was not universal.

The TSE in Camp Mirage was an important part of the TF and a conscious effort was made to include them in all that we did. Once a month, the Commander and I would travel to Camp Mirage for a day. The day normally included operations briefs, a chance for the staff to deal with issues face to face with the Commander and would end with a town hall meeting so the Commander could talk and answer questions from TSE members. It was important that they knew that we appreciated the work that they did and that they understood what it was that we were doing. When the Commander and RSM of TSE came forward to Camp Julien, we would arrange for them to get out with each of the units in order to see what it was that we were doing and under what conditions. After the Change of Command of TSE in June 2005, we continued the same routine with the newly arrived CO and RSM so that they would better understand our requirements. I think that this made for a good working relationship and fostered understanding, trust and respect on both sides of the theatre. For example, the D&S platoon from Camp Mirage came forward a section at a time to reinforce Force Protection Company. In fact, they accumulated enough time in theatre to be eligible for the Campaign Star. Our mutual visits were well worth the time it took to travel back and forth and many commented on how this simple act made them feel included in the bigger team.

We also commenced a programme whereby we visited every unit and sub-unit and participated as a member of that unit. For instance, we travelled with Recce Sqn, spent time in the observation posts, conducted dismounted patrols, gate guard and mounted patrols with FP Company, we went on calls with the IED teams and spent time with the Engineers while they were engaged in ammunition destruction and construction tasks. Furthermore, we travelled with Transport (tpt) on convoys as a tail gunner; and we accompanied the Field Surgical Team (FST) in late night clandestine meetings in the heart of the city which allowed us to see for ourselves if the TF was doing what we thought they were doing. In some cases, we were surprised to find that they were not. At the end of each event we conducted an After Action Review (AAR) and provided direction. The next time we visited, we would confirm that the recommended changes had actually taken place. We attempted to do this as often as possible without becoming a burden to the units.

During this time we commenced a meet and greet program with our CIMIC teams taking the lead. Meetings were set up with the Mayors and Police Chiefs of Police District (PD) 6 and 7, which was the area of the city surrounding the camp. They were hosted in the camp for a meal, which gave us all the chance to meet face to face and hear their concerns and their impressions of us. The mayor of PD 7 was very clear when asked what he thought about us being there: his first response was "if we did not want you here, then you would not be here." He reminded us that they had not wanted the Russians and the Russians were no longer there. He acknowledged that we have provided him and his district with a level of security that he had not known in a long time and for that he was grateful. He wanted us to know that when they, the Afghan Government, could provide the same level of protection, then they would ask us to leave and if we do not exit at that point, then they will make us leave. He cautioned us about our interaction with the local population and warned us that they did not want us to pass on our western ways and ideas in an effort to influence the population. The exchange provided telling insight that was very honest and well appreciated.

Although we promised nothing during these meetings, the gatherings gave us a better understanding of their priorities. The forum provided a unique opportunity to ask questions and provide us with a look at the world through their eyes. After the initial meeting, we met every couple of weeks mostly for social visits or to update each other on projects that we were working on.

Unfortunately, the CIMIC team had no money to use for projects in the area. With a little command influence and the cooperation of the CIDA representative at the Embassy we were provided with a hundred thousand dollars for projects in our local area. These would be small projects, such as the construction of wells or school facilities, of no more than five to ten thousand dollars each and would be tied back to CIDA. We were more than pleased with our new found wealth and presented a project list that was blessed by CIDA. All work was contracted by locals to locals under the supervision of the mayors and other Afghan authorities. This provided the image that the local government was working toward the betterment of life in their area. We benefited from their involvement as well because they found the best prices and ensured that schedules were followed.

This interaction became a theme for the way we conducted business. All too often at the beginning of the tour we had wanted to be the centre of attention when we provided resources for others, even if this is detrimental to the main aim. We soon learned, however, that it was important for local leaders to be respected and even something as simple as the smallest gift should be given to a senior Afghan representative to handout. This cooperation provided a certain level of dignity for all. In the end, we all knew where it had come from but after all we were there to support the government and its institutions.

This cooperation also spawned interaction between us and the medical and police communities. The medical staff started to pay visits to the local hospital to provide whatever help they could. The military police began a training program for the Kabul city police from PD 6 and 7. The Afghan police at this time received no formal training. The military police (MP) and Force Protection Company gave them training in weapons handling, basic search and vehicle search procedures, as well as vehicle check point setup and operation.

These meetings actually paid off as the tour continued. The locals began to pass information about possible operations or dangers in our area.

THE THREAT AND ITS IMPACT

This mission was unlike anything that most of the soldiers had been on before. There were no opposing forces with established orders of battle (ORBATS), no front lines, no declared zone of separation, no checkpoints and no uniforms by which to identify the threat. The threat could be anyone, anywhere at anytime. It was possible that the threat was represented by something different to each individual in the TF. With no clear picture of the threat, individuals could conjure up all kinds of demons. The first trip through the unfamiliar city was no doubt stressful as you watched every person, vehicle, donkey, camel, and every piece of garbage left on the side of the road. You soon began to realize, however, that until you have spent enough time travelling the area to know what is normal, you could not distinguish what was a threat and thus remained at the mercy of luck. Those that did not or could not come to grips with this reality suffered greatly, some to the point where they were under continuous stress.

During our visits to some of the units it became apparent that some of the intelligence information had been inflated or misunderstood leaving some with the impression of extremely high threat levels. Rather, this information should have been used to mitigate the risks and plan appropriate actions. Some units lost sight of the primary mission and instead remained in search of the elusive white Toyota filled with explosives or the possible suicide bomber. These units needed to be refocused on their primary mission. The threat needed to be put in perspective. For instance, during one incident, the Commander's vehicle was almost side swiped by two of our own vehicles from Transport Platoon while exiting a NATO compound in the city. The two vehicles then pulled an about turn in the middle of the road and sped away in the opposite direction. Upon investigation and questioning of the personnel about the incident, they explained that they had seen a suspicious vehicle on their way into the compound. They had then left the secure compound in order to follow the vehicle and consequently nearly caused two accidents. We asked them a few more questions. First, why did you think the vehicle was suspicious and what were you going to do if you caught the vehicle? Second, why did you leave a secure compound that had a mounted QRF ready to react to such incidents? They had no answers. They had clearly misunderstood what their mission was, what was expected of them, and how to deal with the situation. All members must be aware of their surroundings and be able to report things that are out of place. Not all are trained, equipped, prepared, or authorized to take unsupported direct action, however.

The overall threat situation in February was quiet. The heavy snow in the mountains had made travel difficult. There was an uneasy calmness in the entire country.

In the first weeks of the mission the All Source Intelligence Center (ASIC) informed us that a foreign intelligence agency had received information from a very reliable source that a suicide bomber would attack Camp Julien. This was the first time to our knowledge that the camp had actually been named as a target. This set into motion a series of events that highlighted some weaknesses in our TF and equipment.

A heightened level of security was set in place. Security checks were established for all local employees. The security at the gates would have to pay closer attention for explosives. As we quickly found out, this was not as simple as it appeared on paper.

It soon became apparent that we had no reliable method by which to detect explosives. The gate had been using the detection swabs similar to those used in airport security. These had been used so many times that they were of no use and there were no new ones in theatre. They had continued to use the ones that they had as a deterrent, hoping that no one would notice. It became imperative that the TF find a solution to this problem. Many of the NATO Forces used dogs trained to detect explosives. The Norwegians allowed us to use their dogs and handlers to do a sweep of the camp and then for periodic use at the gates. The main threat was the walk-ins. They were mostly locals who worked for us and the food truck as it was the only outside vehicle that came into camp. All locals were searched closer than normal and this became a point of discontent as most of them had been with the Canadians since the camp had first opened. We then embarked on a project to get our own dogs and handlers as well as to secure some electronic methods of detecting explosives. A company in country could provide the dogs and handlers and a contract was signed with them. The explosive vapour detectors used by most security firms were bought later in the tour. They arrived in theatre in May. Unfortunately, they had not been calibrated for the altitude of Kabul and would not function until someone from Canada came to recalibrate them.

The NSE and Engineer Sqn mounted extra platoons that were given to FP Company to provided additional personnel for security. Recce Sqn was still conducting Op Horseshoe and had no additional personnel for this tasking. The D&S platoon in Camp Mirage sent an additional section forward. Given the size of the camp and the number of positions to be manned around the clock, the TF slowly ground to a halt. Vehicle on repair rates rose and day to day tasks began to suffer.

After nine days, the TF could no longer sustain the level of activity that it had inflicted on itself. The TF was tired and the level of security was consequently reduced to a manageable level. The incident provided us with insight on what the capabilities of the TF were; with all its amenities and because of its size, the camp was the problem.

This was not the last occasion that additional personnel were required for security. There was a constant threat of rocket, mortar, missile and suicide bomber attacks against the camp.

It became apparent that we could easily become overwhelmed with the amount of intelligence that we received. We had to develop a method to filter or qualify what it was truly telling us. After some contemplation, it was decided that we needed to place this information in context. We did this by profiling previous attacks in our area, in addition to all of Afghanistan, and some attacks that had happened in Iraq.

The threat of rocket attack turned out to be real and happened numerous times in the city. Recce Sqn spent a great deal of time trying to disrupt and detect these attacks. When we began to profile the attacks we discovered the most likely location based on range, time of day or night, moon or ambient light conditions and the average number of rockets fired during an attack and the interval between rockets. This information allowed us to ramp-up for shorter periods of time if optimal conditions for an attack existed. Some of the other events were a little more difficult to profile but this still provided a better picture of how or what to do to mitigate our response to the threat.

There is no way to qualify or quantify our success or failure. It is much like everything in the intelligence world and one needs to ponder whether having the information makes you actually safer. Was there a real threat? Have we actually defeated or disrupted the threat? Having said that, we were wiser for the exercise, we did respond to each threat and, as time went on, we could better manage our responses.

Each incident presented a number of options regarding how to handle the threat situation and created a good deal of discussion about the best course of action. The Commander and I at times had differing views as to how we should deal with the situation. I perhaps had a more cavalier attitude hoping to break the Fort Apache attitude. I felt that when there was a heightened threat, we should project ourselves in force. He, on the other hand, had a due diligence approach – which was understandable given the situation. I felt that we needed to break the "turtle attitude" and not only the Commander, but the system as a whole, had to accept the risk that this entailed. In the end, the due diligence approach won. We knew that no one back home, at least then, was prepared to accept increased risk. Nonetheless, I give the Commander credit for having listened to me each day until a final course of action was firmly decided upon. Once the decision was made, then that was the end of the discussion. This is a lesson that some NCOs needed, and still need, to learn.

March brought some relief from the harshest winter that the country had seen in decades. It was reported that thousands had died from the conditions and tens of thousands were left near starvation. Due to these factors, it became apparent that the election that had been scheduled for April would not happen. The winter weather and lack of preparation time meant that the government had to delay the date until September.

At this time, we were informed that Canada had made the decision that Kandahar would be the location for the PRT. It would be part of the next rotation. The newly designated Deputy Chief of the Defence Staff (DCDS), General Dumais, was coming to see the theatre and requested a meeting with the American TF and the PRT Command. We flew back to Dubai, met him and then accompanied him to Kandahar. He visited the TF, Embassy and higher HQs in Kabul before returning home.

The first week of March, a UN worker, Mr. Steve McQueen, was shot and killed on a Kabul street. Recce Sqn was conducting operations with the

German Battle Group in the Surobi area east of Kabul when a patrol was engaged with small arms fire. There were no injuries and no one was captured.

The third week of March, four American soldiers who were working with the ETT training the Afghan National Army (ANA) in Kabul were killed in a mine explosion in Logar, just south of Kabul. They requested our Engineer Sqn to conduct a forensic investigation to see if they had been targeted. Sadly, they had just wandered into an unknown minefield. Having members with the ETT, we attended a very moving memorial service with the American unit.

Near the end of the month, one of the Embassy vehicles was destroyed by an IED on the Jalalabad road, designated as the Violet Route in Kabul. This route ran from the city centre east out of the city to Jalalabad and then to the Kyber pass. In the capital region, it was the main transportation link to Pakistan. This route had been, and would continue to be, the site of many attacks against coalition forces in Kabul. Luckily, the Ambassador was not in the vehicle at the time of the attack. The bomber was captured running away from the site of the attack and throwing his backpack into the river. From interviews conducted later we learned that he had not targeted that specific vehicle. An Iranian "handler" had trained him in Iran. His handler was located some 200 meters from him on the route where the bomb was detonated. He had been targeting any western style military or civilian vehicle. On three occasions he had been ordered to detonate the bomb but had refused because it would have killed too many locals. Finally, he was told that if he did not detonate it when the next western vehicle passed, his handler would kill him. The bomb consisted of 11.5 kilograms of explosive in a wheelbarrow on the side of the road. The detonator was activated by remote a short distance away. The blast took out all four tires, cracked the engine block, and blew out the rear non-armoured windows of the vehicle. It drove pieces of the wheelbarrow through the outer skin on the doors, through the body and nearly took the passenger window out of the frame. The armour protection in the passenger compartment saved the driver and the military police security personnel. The two were disoriented and scared but were not severely injured. Approaching the area from the opposite direction were three members of the Canadian ETT who quickly secured the location against secondary attacks and provided medical attention. This was the beginning of the "spring offensive" in the city and would be followed by many more attacks.

April saw an increase in belligerent activity throughout the entire country. The Americans lost a Chinook in a dust storm just south of Kabul with the loss of 18 soldiers. In the city, two Kabul policemen were shot and killed by the Taliban. A bomb mounted on a bicycle and left on the side of the Violet Route took the rear windows out of a NATO vehicle and destroyed a taxi that was following it. The taxi had five occupants and all sustained serious wounds. In the eastern part of the country, firefights along the borders were becoming a daily occurrence. In the south, clashes between poppy farmers and US and ANA troops ended in a number of deaths on both sides.

April saw Recce and KMNB conduct an operation west of Kabul in Pagman. This was an operation to gain information about a known warlord, weapons dealer and former Taliban commander. After initial positive results, within about nine days, the mission was shifted to another area. There was confusion concerning this quick withdrawal while gains were still being made. Some questioned interference from the outside, but these concerns were never confirmed. The month also brought with it more threats of rocket attacks. An American working for one of the non-governmental organizations (NGOs) in Kabul was kidnapped and placed in the trunk of a car. While driving down one of the main routes in Kabul he managed to get loose and jumped from the moving vehicle where he was helped by some of the locals.

May brought a suicide bomber to the Kabul Café killing five people and wounding two others. Moreover, an Italian aid worker was kidnapped in broad daylight in the middle of the city. She was later recovered. Some say that a ransom had been paid. Rockets were once again hitting the capital with ISAF HQ and Embassy row being hit, luckily, with no casualties. Just north of the city a house containing an unknown quantity of ammunition exploded killing twenty-eight locals. The Germans suffered a casualty in a mine strike just east of the city on a well-travelled route. The eastern and southern parts of the country also saw a great deal of action. In one incident it was reported that an Afghan patrol was fired upon and returned fire. The Afghan forces asked for and received reinforcements only to find that the enemy force had also grown. The firefight raged for three days with the employment of all the assets that the Americans had available. In the end, more than 150 insurgents were left dead in the field. This was of great concern because it was previously believed that the Taliban could not muster those kinds of numbers.

June arrived with a warning for General Wardack, the Afghan Defence Minister. The insurgents were changing their tactics and were grouping in units of 60 to 150 insurgents and would most likely attempt direct ambushes against coalition forces. Attacks intensified throughout the country in an effort to disrupt the selection of candidates for the September elections. In the first week of June, an IED is found just outside the Camp and disarmed by our Engineers. An IED was set off in the city every four or five days. Most created little physical damage. Two German soldiers were killed when something went wrong during an ammunitions recovery operation. Reports of up to 60 armed men moving through the mountains from Logar province had Recce deployed into the mountains monitoring all passes spread across a 60-kilometre front, leaving only a few patrols left behind to cover the capital city. The Recce Sqn had been going full-out since the mission had commenced. One of the Recce operations in the city was targeted and received rocket fire, just missing the site by one hundred or so metres. The British QRF responded but no one was captured. Additional covert surveillance was established on the launch site in hopes of capturing those involved.

July saw some gains as the covert operation found three more rockets aimed at the operation, and this led to the capture of those responsible. The Slovenian troops attached to Recce struck a mine when one of their Hummers slid off the road. The mine destroyed the Hummer removing the entire rear end. One of the crew suffered ruptured eardrums but all were able to walk away. There were three more rocket attacks into the city. A Vehicle Borne Improvised Explosive Device (VBIED) killed two Kabul police and three civilians. The Americans working in Kohst province near the eastern border, had a special forces (SF) team come under fire and requested assistance. Chinooks and Apaches from Bagram were dispatched. Flying between the mountains in a dust storm the lead Chinook was downed by rocket propelled grenade (RPG) fire. All on board were lost. One of the four members of the SF team escaped and was later extracted.

One would have thought that the TF would have become withdrawn because of all of the attacks but the opposite was true. Most had become comfortable with their jobs and gained confidence in their capabilities. They knew what to look for and how to avoid or react to the situation. They had accepted the fact that there was only so much that they could do. The rest was out of their control. This acceptance showed in their calm, measured approach to their tasks. Of course this also indicated one other thing: we were about to hand over to the next rotation. We wished that we could have passed on our personal experiences as part of the handover but you truly cannot and the next team would have to learn the same lessons for themselves.

August brought more of the same but the next task force would have to deal with the TOCA that was set for 04 August 2005.

THE CHANGING FACES OF THE MISSION

As the TF headed out the door, there was the knowledge that things were about to change. The military was looking at shifting mission areas and establishing a PRT somewhere in Afghanistan.

As early as the beginning of March, with the visit of the DCDS, it had become clear that Kandahar would become the future home of the Canadian contribution to Afghanistan. With this knowledge, requests for recces of the route to Kandahar were sent. They were not authorized, however. The route was one of the more dangerous routes in theatre. Regardless, eventually we would have to start shifting stores and equipment south. There had been a 250-person camp already loaded into sea containers in anticipation of the requirements for the PRT.

The Chief of the Defence Staff (CDS), General Rick Hillier, visited in April. The troops hoped a formal announcement would accompany him and allow for some movement south. This was not to be; the Canadian Government had not signed off on the new mission. There was conversation, however, about a new element called the Strategic Planning Team, an organization to help the Afghan government with strategic level planning. April also brought the tactical recce for the PRT and the TAT for the Kandahar site. Still, however, there was no authorization to commence the move south.

April also brought the Tactical Recce for Roto 4 into theatre. The new TF had a different structure than ours and there were two CWOs taking over my role. I spent the better part of five days with them. We travelled the AO and I introduced them to all the people they would be interacting with. I passed on as many lessons learned as I could. Before departing, they discussed the time required to complete the relief in place during the actual rotation. They had indicated that perhaps five days would be required and I protested. I told them that we would require no more than two days. I instructed them, "when you arrive I will brief you on anything that has changed since the Recce and any legacy issues that you may have to deal with. Following that, I will send my last sitrep [situation report] to the DCDS CWO and then sign off the computer and vacate the chair." Following which point I told them, they would be in charge.

May brought the official government announcement that the mission would move to Kandahar after the election results had been finalized. A camp close out team from Canada arrived to develop the plan for the camp teardown. This announcement also allowed the route recce to commence, followed by the movement of stores to Kandahar.

The TAT arrived in Kandahar in June to prepare the site for the PRT. The Strategic Planning Team recce arrived and accommodations and offices were secured for them. The Commander and I paid a visit to Kandahar to see the TAT. The TSE change of command was in early June, in advance of the TF rotation.

July saw the arrival of the PRT. The TAT continued to work on the Kandahar site. The TF started to arrive in late July. Camp tear down had already commenced in advance of the rotation.

All of these tasks were completed with no increase in staff for the TF. This no doubt placed an extra burden on the staff and members of the TF. At the end of the tour, the TF was tired and ready to leave but we were confident that we had set up the next rotation for success.

REDEPLOYMENT

The redeployment plan was something that was considered once the TF settled into a routine and home leave travel allowance (HLTA) had commenced. There would be a plan developed that would encompass both Regulars serving with 2 CMBG and Augmentees and Reserves. Most of the medical and social worker aspects were taken care of in theatre, especially for those from outside the Brigade.

The TF would stage through TSE, most with a day or so in Dubai. There were a number of activities that they could participate in while waiting. Some consider this stage a necessity and even what the soldiers want. From my experience, the soldiers just want to go home. For some reason, however, we have convinced ourselves that if we do not provide some sort of decompression period, we will have failed our troops. Personally, I too just wanted to go home. There was a need for a break in Dubai because there was a maximum of two Hercules loads each day that could be brought out of Kabul. The Airbus could hold at least three Hercules loads, so in order to fill it, the day layover was required. This was interrupted when an aircraft damaged the runway in Dubai making it unusable for four days. The flights were changed and some transited directly from the Hercules to the Airbus without a layover.

Those that arrived in Trenton and were not part of the Brigade had to remain overnight and then go through a series of medical, padre and social worker checks before being allowed to leave. I found this added delay was very irritating, as all of the things that were done there had already been completed in theatre. Moreover, as a Regular force member, I could have done this at my home unit.

MAINTAINING DISCIPLINE

There were few disciplinary incidents during the tour. There were only twenty-five charges in Kabul, including seven for Negligent Discharges (ND). TSE saw only seven charges in total. The TAT had one charge for an ND. Most other charges were for housekeeping. With the exception of the ND charges, the others were nothing more than any unit of the same size would expect in Canada. This may have been an indication of how busy things were. Most members went from their work area to the mess hall to the gym and then to their quarters.

The NDs that were experienced were, for the most part, caused by a lack of attention or fatigue. Most were fired into the unloading bays. One in particular was fired by a Senior NCO after arriving back from HLTA. The trip back into theatre was long, taking almost three days. The ND was fired through two portable toilets into the kitchen area passing through a mess tent during supper hour. It ended up in the food preparation area, where it was embedded in a pot being carried by one of the staff. It was a very close call and to this day I have no idea how someone was not hit by the projectile. The lesson here was to pay particular attention to those that have just arrived from HLTA; after almost three days of travelling they are not competent or coherent without a rest period in theatre.

QUALITY OF LIFE

As was mentioned, the camp initially appeared to provide a large array of distractions and activities. It did not take long for the shine to wear off, however. The TF was very busy during the entire tour and for those that could not leave the camp, it became a prison. At times this sparked resentment of those who spent most of their time outside the wire. Upon occasion there was a need to remind all of what our focus, direction and mission were. With no R&R and a no walking out policy, work was the only distraction. The HLTA was a welcome break for all. As the facilities in the camp started to come down, there were many more complaints. The first things to be taken down were the common areas, games rooms and some messes. This coupled with the increase of 400 extra personnel being brought into the camp for the build up to the elections left it very crowded.

Although we did not have any casualties from hostile attacks, we did have a number of individuals that had to be medically evacuated. One individual had reported to the hospital complaining of a headache and generally not feeling well. The medical staff decided to hold him in the hospital overnight for observation. During the night he left his bed to use the washroom and collapsed. His system had shut down and he needed to be evacuated to the role three hospital in Germany. After a number of days there, and without a clear diagnosis, he was sent back to Canada. At home he was diagnosed, treated and returned to theatre within a month. Each incident provided us with the opportunity to test and refine our procedures without having to take casualties to do it. Nonetheless, for various reasons, there were fifty-four personnel repatriated during the tour.

TRAINING AND EDUCATION

Training continued throughout the entire tour. Some training included all of the TF, while other training was targeted at specific individuals or groups.

The Assistant Judge Advocate General (AJAG) had put a package together to ensure that all units were dealing with disciplinary investigations and the prosecution of charges in the same manner. This was delivered to Senior NCOs as well as officers. The units conducted Tests of Elementary Training (TOETS) on a weekly basis and had their own schedule of professional development for both NCOs and officers. Each month a competition was scheduled based on the skills that a soldier should possess, for example, shooting, first aid and navigation. It was as much for fun and a distraction from the day-to-day routine as it was to perfect the soldiers' skills. Ammunition for ranges was allotted and most had the opportunity to fire each month. The mission's high tempo sometimes limited the participation in these events but nevertheless the training seemed to have been well received.

The education of the TF was dependant on the level of command or responsibility. There were professional development (PD) sessions on the history of Afghanistan from the British era to present day. Part of the NCO education was to try and have them develop a critical eye toward the details of the operation. This required them to ask if our operations are effective and efficient or if changes were needed based on the threat or any other factors. This exercise met with some success, but there seemed to be reluctance to change those things which seemed to work, regardless of whether or not it made sense to continue with them.

CONCLUDING REMARKS

By the time the mission ended, I had reached a level of comfort with the personnel. I wished I had been able to obtain this before entering the theatre. That being said, I could only hope that we had all learned something during this time.

The TF functioned well and, for the most part, the training had been effective. It is always our desire to make it better but we all know this must be measured with the cost of extra time and resources. For example, we should have embedded teams with each rotation that will be there to gather lessons learned and to provide training scenarios for the following task forces. This would hopefully give an avenue for all lessons learned to be funneled through and would ensure that none are lost.

It is not possible to replicate the capabilities of the theatre during the lead up training or during normal training. This will continue to be a downfall of our system. We cannot wait until we are deployed to see or have access to the range of resources that are presented to us in theatre.

There were perhaps many things that could have been done differently but in the final analysis it worked. All took away from the mission their own lessons learned, both good and bad. Our success is a testament to the adaptability of our soldiers and to a concerned leadership. I am glad that I had the opportunity to go into theatre as I believe that this is the real reason that we, as Canadian Forces members, are being paid.

CHAPTER 13

TASK FORCE AFGHANISTAN 2006

Kit Charlebois

I was the Task Force Sergeant-Major for Operation (Op) Athena, Rotation (Roto) 3 also known as TF 1-07. During my deployment we faced many challenges.

One of the unique challenges for this specific Task Force (TF) was that the majority of the troops were force generated from Land Forces Atlantic Area (LFAA). Most of the Reserve soldiers that deployed with this TF were from a variety of Reserve units from across the Maritimes. This type of force structure (i.e. mix of troops) had never been done before to the best of my knowledge.

The other major challenge that we faced was that many of the enablers (i.e. combat support functions) attached to the Battle Group came from across Canada. For example, the Reconnaissance (Recce) Squadron and the Artillery Battery came from the Royal Canadian Dragoons (RCD) and the Royal Canadian Horse Artillery (RCHA), respectively, out of Petawawa. The tank squadron and the 3rd Rifle Company (Coy) came from Lord Strathcona's Horse (LDSH). The core of the Battle Group of course was based on 3 Princess Patricia's Light Infantry (PPCLI) out of Edmonton. Other support personnel also came from across the country.

Certainly, this was a diverse TF composed of soldiers, airmen and women and naval personnel from one end of the country to the other. Consequently, preparing the TF for the road to high readiness was a huge administrative challenge and put all levels of leadership to the test.

For example, even during our confirmatory validation exercise in Wainwright, Alberta we were still trying to identify what vehicles with which the Light Infantry Company out of 3 PPCLI were going to deploy. The lengthy process created frustration for the Battle Group, the Light Infantry Company, the National Security Element (NSE) and the National Command Element (NCE).

Additionally, we had to keep the Reservists focused. The majority of them had already been training for six months away from home prior to joining the Battle group in June and then the remainder of the TF in September. In total, the reserve individual and collective training, the TF collective training and the tour was an eighteen month experience. Many of them did not live near training bases and therefore did not get to their families much.

In the end, the mounting of the TF out of LFAA was a success. This achievement was due to all personnel in the TF and the force generating unit's support that had one focus in mind: to prepare all personnel in the TF for battle.

On arrival in theatre, there was not much downtime as the Battle Group immediately commenced indoctrination training and started moving out to the Strong Points and the Forward Operating Bases (FOBs). Our first task was to get to know the Area of Operation (AO). All levels of leadership started to engage with key local leaders and got to know the lay of the land through numerous foot patrols.

On one of my visits to one of the strong points in the Zahari-Panjwayi district, the platoon Warrant Officer (WO) informed me that there was a section departing on a night patrol. Due to the intense rain that had occurred during the previous few weeks, he was concerned about the overfilled wells in the area. In an effort to solve the problem, a mountain operation instructor who was a member of the platoon set forth and jerry-rigged a number of empty 60 mm mortar tubes. He cut a hole in each end and then proceeded to coil approx 100 meters of boat rope into each tube. He completed the process by closing the tube and tying a knot at each end. He made approximately nine of these devices for each of the patrols that were to exit the wire. This device was designed so that if a patrol member fell in a well at night, then a patrol member carrying one of these devices could take the cap off one end and throw the mortar tube down the well to the soldier and the patrol could then pull him up. Notably, this example underscores the variety of skill sets that a non-commissioned officer (NCO) acquires throughout his career. Without a doubt, it is important to keep sending our soldiers on courses that develop their skill sets.

We arrived in Afghanistan during the poppy harvesting season thus Taliban activity was at a minimum. This lull gave the TF the opportunity to show its presence through the complete AO. The temperature was hot and the battle rattle was quite heavy. The troops did not complain, however. Moreover, they knew that it would only get worse as summer was fast approaching.

Our first fatality happened shortly after our arrival in theatre. There was a shooting in Kandahar Airfield (KAF). The soldier who died from his wounds was part of the force protection Platoon that secured the front entrance to KAF. This platoon was made up entirely of reservists. This sad occasion marked the first time that I personally noticed how close the reservists were to each other. The death of the soldier was a major shock to the platoon. Many of them were very close because they had been soldiering together in their home units for decades. They felt the loss akin to losing a brother. Their closeness proved to be a double-edged sword, however. Unfortunately, one of the soldiers called the fallen soldier's family prior to the communications lockdown and informed them of their son's accident. This action did not help the Assisting Officer who went to inform the family of their son's death. When he got to the residence it was full of angry relatives looking for answers. The TF learned from this experience and tightened up the communication lockdown procedures

One of the first things that I changed in theatre was the ramp ceremony. I established a viewing area for fallen soldiers should we experience any fatalities. Having dealt with many military funerals before, I fully understand how important closure is for the troops.

During the ramp ceremonies on the previous Roto, the pallbearers marched the transfer cases onto the C-130 Hercules aircraft for their departure home and on dismissal of the parade everyone clambered onto the C-130 to pay their last respects. I personally witnessed this ceremony when I travelled to theatre on the Tactical (Tac) Recce. During our time there with Roto 2, the TF lost three soldiers to a suicide bomber. Consequently, I had the opportunity to witness the ramp ceremonies in KAF, with the Theatre Support Element (TSE) and in Trenton, Ontario. What concerned me in KAF was that with everyone climbing on the back of the C-130, the ceremony became a media event. Moreover, there was no appropriate place to address the crowd as they clamoured in the back of the plane.

After discussing my plans to change the ramp ceremony with the TF Commander and receiving his concurrence, I brought the Company Sergeant-Majors (CSMs) in and explained to them my intent so that they could brief their soldiers. From that point on, when we had a fatality, or fatalities, the transfer cases were marched on and the pallbearers marched off the aircraft and no one was allowed back onto the aircraft. Additionally, I approached the mortuary affairs in KAF, which was run by the Americans, and explained to them that I would like to set up a viewing area in their location as it was the only fenced off area close to the Roll 3 hospital that would afford a secluded area large enough to enable the TF the opportunity to pay their last respects. The supervisor of mortuary affairs was a National Guard Sergeant whose job back home was as a funeral director – he understood my requirements.

It was not long after that reorganization that we received our first fatality, the shooting at KAF mentioned earlier in this text. My staff from the NSE, Commanded by a Royal Canadian Regiment (RCR) Master Warrant Officer (MWO), set up the viewing area with the NCE. The area was lined with white camouflage netting with the Canadian flag draped behind the transfer cases and the transfer cases positioned side by side with a small table beside each one that displayed the fallen soldier's beret and an 8x10 photo. When the TF came by to pay their last respects I allotted the first thirty minutes for the sub-unit (i.e. company (Coy)) of the fallen soldier to have their own time with their fallen comrade before the remainder of the TF was allowed to pay their last respects. What I noticed at these viewings were many people breaking down and hugging each other in groups and certain individuals just sitting beside the transfer cases and reflecting. I also noticed that many of the soldiers would place small mementoes on the table such as pictures, metal wings or a meaningful personal item.

This new viewing venue and ramp ceremony proved itself time and time again. I received positive feedback from the TF that this provided the right closure for the troops and that they could now go back to the area of operations (AO) and focus on their missions. Moreover, the TF padre informed me that the mental health workers in KAF were no longer seeing the usual number of patients that they had become used to following a loss in the TF.

I also noticed that when we lost someone during the tour, it affected everyone in the TF as we had all became a close family in battle. Everyone knew each other in one form or another and it was important that the opportunity for closure was afforded to everyone in the TF.

On my tour we lost twenty-two soldiers. Unquestionably, this was a huge loss. At one time I lost seven warriors, six of whom were soldiers and one was an interpreter. At their ramp ceremony, a couple of senior officers approached me and commented that they had not understood the true magnitude of the loss until they saw the forty-eight pallbearers and the pallbearer commander marching out of the C-130 after the lowering of the transfer cases.

Certainly, a huge challenge for the commanders at all levels was our loss of soldiers during the tour. We incurred fatalities due to mine strikes and Vehicle Borne Improvised Explosive Devices (VBIED). Consequently, it was a challenge to keep the troops focused on the mission. Everyone knew the Taliban's tactics had changed since Op Medusa and they were now hitting the TF with numerous Improvised Explosive Device (IED) strikes of all forms. Everyone who went outside the wire was vigilant in his or her duties and paid extremely close attention to their orders and their specific tasks. Not one person was exempt from the dangers of IED strikes and when departing the FOB or KAF the soldiers were focused and prepared for any eventuality.

As the seasons changed, so did the Taliban's tactics. Once the poppies were harvested, the Taliban commenced low level tactics of hitting the Battle Group with section to Platoon sized insurgent groups. The Battle Group was extremely successful in defeating these attacks and in the case of "I" Coy, most of the time they took the fight to the enemy. One significant point to mention is that to the rifle Coys and the enablers to the fighting echelon troops, this was a young man's war. With the 55 degree Celsius heat that we were experiencing and the heavy loads that were carried by all personnel in their battle rattle it was a leadership challenge to ensure that the troops were hydrated at all times and that operations were planned in consideration of these extreme temperatures. During the pre-deployment training many of the rifle companies focused on weight load marching which paid dividends overseas. The CSMs had to constantly monitor the troops in theatre to ensure that they were eating properly and that they were taking their electrolytes and hydrating themselves prior to an operation. The company medics kept an eye on the soldiers' diet and informed the chain of command (CoC) as to whether or not the soldiers were eating the necessary supplements to keep them sustained in operations.

Nonetheless, we suffered many injuries, including sprained ankles, slipped discs and dehydration. Some of the injured were repatriated back to Canada.

Our TF was dispersed throughout the AO with the companies inheriting assigned districts with the aim to dominate the ground and understand the culture of the locals in their AO. This included the Recce Squadron and the Artillery Battery who worked out of their respective areas.

Being dispersed across Regional Command (RC) South, it became a challenge for the NSE to re-supply the TF. This is how the CSMs, Company Quartermasters (CQs) and second in-commands (2ICs) really earned their money. Because of the age and time in of our CSMs and CQs, and the experience that they carried, the skills they used to facilitate the re-supply were derived from what they had learned during their Warrant Officer (WO) courses way back in the 1980s and 1990s. The CQs completed the demands from KAF and sent them with the Combat Logistics Patrols (CLP) from the NSE to an interim location where the CSMs or the Stores 2ICs would meet the CLPs in the FOBs and transfer the loads to their vehicles. They then prepared to move to the forward troop locations throughout the AO. In the end, this time tested manner proved to be the best way to accomplish the replenishments.

Due to the austere FOBs that the Battle Group lived out of, the CSMs once again had an opportunity to put their previous training to good use. In Gundy Ghar, a FOB in the west of the TF's AO, there were no ablutions as it was situated on a barren hill that dominated the ground for many kilometres in all directions. The ground was pure moon dust. The CSM had to educate the Squadron on how to burn the excrement in the mornings as there was no other way of removing the waste from the FOB. He had to get his squadron to build outhouses out of spare wood as well as construct field showers. The ingenuity of all ranks was very apparent in the construction of these temporary ablutions which provided some comfort in the harsh conditions.

The troops did not have a problem digging in and building bunkers to withstand rocket attacks, which occurred quite often in the FOBs. Keeping weapons and equipment clean was a never-ending task. Each FOB had unique challenges. Once in the FOBs, not only did the CSM have to get the troops ready for the next day's battle but they also had to be the Camp CSM.

Home Leave Travel Allowance (HLTA) periods took their toll on tour as they often resulted in a leadership vacuum; however, they also enabled the corporals the opportunity to step up into the section 2IC spot and the master corporals to be raised to section commanders and so on. Once again, the excellent training that our soldiers had received prior to deployment was evident in their ability to perform beyond their rank. This high calibre of training was evident when one of the platoon 2ICs stepped up in the Platoon Commander's spot during the Platoon Commanders HLTA and led the platoon successfully in battle.

During the tour I constantly monitored the dress and deportment of the soldiers. The non-commissioned member (NCM) leadership, starting with the CSMs down to the master corporals, ensured that the troops followed appropriate dress regulations. Prior to the TF officially forming in September, I had brought in all the CSMs and sat them down and discussed the issues that pertained to the TF. We brainstormed on all the issues that we thought we might be faced with overseas, covering everything from dress, discipline, deportment, ramp ceremonies, morale and health just to mention a few. Once we established the "dos" and "don'ts," I briefed the Commander on our conclusions and subsequently we put these regulations in place. By setting the standards of the TF early and having the full support of the CSMs, the dress and deportment was always kept in check and we all looked like one fighting unit.

One of the other challenges that we had was the discrepancies that arose in the different AOs. For example, one of the companies covered a sector that was relatively quiet of Taliban activity and they became battle envious. Conversely, a different sub-unit was fighting the Taliban on a daily basis and getting weary of the challenge.

The AO was dynamic and represented a three block war. In one sector there was full out battle with all the enablers, in another sector there was reconstruction headed by the Provincial Reconstruction Team (PRT) and in yet another sector the Governance piece was being played out. All of these sectors were relatively close to each other. My Commander had his plate full trying to balance all three of these demands that were placed upon him. This reality made for a very diverse AO.

I felt that we had gotten it right with regard to the training that the TF had gone through on the road to high readiness. Albeit, it did take many folks away from their loved ones for long stints at a time but, in the end, it paid dividends. Every Commander made sure that everyone going over knew their Tactics, Techniques and Procedures (TTPs) inside and out.

I was hit by a VBIED five days before departing theatre and if it was not for the trained crew, who immediately put their drills into action, I would not be here writing this today. Although shaken up, all of us exited our vehicle and commenced our post-blast drills of establishing the inner and outer cordon, assigning arcs of fire and securing the blast site. Thinking back on that day, I realize how important the pre-deployment training is. Anyone who will be going outside the wire and has not been put through any type of Theatre Mission Specific Training (TMST) to deal with events that could occur outside the wire is a liability to the TF. Once outside the wire, everyone has a role whether that is a medic, a shooter, crew commander, gunner or driver to mention but a few.

The visitors that we received caused tension and frustration within the TF personnel because of the extra requirements it took to protect them. Everyone was vulnerable and susceptive to attack upon exiting the wire and trying to fight the enemy while simultaneously protecting the visitors was forefront in the minds of the men and women in the TF at all times. Unfortunately, it will take a serious injury to a visitor to drive the point home that Afghanistan is not the place for military tourism.

Another challenge that arose toward the end of the tour because of the loss of the twenty-two men was the lack of motivation among some of the engineers. Most of them had been at every catastrophic blast site and they had recovered both bodies and vehicles, participated in mine clearing and/or provided local security. After a number of ramp ceremonies, increasing numbers of engineers could not exit the wire as they were battle fatigued. Consequently, the Battle Group had to put different engineer sections together in order to provide engineer support on various missions.

All in all, Roto 3 was a success because of sound leadership at all levels and of finding the right balance of grit and compassion. I have a new appreciation for the Reserves as they put themselves in harm's way as many times as their regular force counterparts. In fact, during the tour, based on the way they fought, I could not distinguish reservists from regular force troops. The only distinguishing factor was the berets they wore at the ramp ceremonies or on medal parades.

Despite our successes, there was room for improvement. One of the changes that I recommend is to insist that the senior leadership take their HLTA in the middle of the tour. Many of them chose to hold off until the end so as to facilitate troops who wanted a middle block. However, this created serious fatigue among some senior leaders. While their intentions were good, you could clearly see that towards the end of the tour, they were tired and close to burning out. Working seventeen hours a day at high tempo in a battle situation will take its toll and affect your state of mind. I witnessed burnout at all levels and noticed that trying to command, at any level, without the proper rest and diet will diminish effectiveness in leading the troops. Those who took their HLTA in the middle of the tour regenerated their batteries and were fresh to command their troops on their return and, importantly, stayed that way for the remainder of the tour.

It was quite obvious to the whole TF overseas that the Senior NCO Corps was the backbone of the mission. Indeed, many of the operations were conducted at section and platoon levels. Striking the proper balance between grit and compassion at all levels was essential. One of the WOs from the TF lost nine of his men, including his platoon commander, to mine strikes and was able to keep his men focused on the mission despite the enormous loss to the platoon. The Battle Group Commander decided to keep him in the platoon commander's position rather than request a replacement platoon commander from Canada.

It was never more apparent than in battle that we needed to lead by example. Those who did not live up to this requirement were replaced and either sent back to Canada or sent to a staff job in the TF. It was a proven fact that the lack of force-on-force fatalities and injuries were due to sound leadership at all levels and an intelligent knowledge of tactics.

At times the Battle Group thought that they were being hard done by by the Commander but, after working closely with the Commander, I can attest that nothing could be further from the truth. He had the Battle Group's interests first and foremost in his mind at all times and he shared in their pain when there were fatalities and injuries. Nonetheless, every organization thought that the Commander was hard on them. That was simply his way of doing business and getting results, however. When I had the CSMs together once a week I would always remind them that the Commander had their organization's best interests at heart. The Deputy Commander was a very compassionate leader and was a good offset to the Commander.

In conclusion, I must say that I was extremely proud of the TF and it was an honour to be their RSM. I could not have asked for a better sergeant-major team. Their leadership saved lives and kept the troops focused. They were constantly outside the wire sharing in the risks of their soldiers. Additionally, the TF was blessed with an outstanding commander and deputy commander.

CHAPTER 14

ANIMOSITY IN THE RANKS: THE DIFFICULTY IN INTEGRATING RESERVE AUGMENTEES INTO REGULAR FORCE BATTALIONS

Kurt Grant

> *The fundamental cause of the breakdown of morale and discipline within the Army usually comes to this, that a commander or his subordinates transgresses by treating men as if they were children or serfs instead of showing respect for their adulthood.*[1]
>
> S.L.A. Marshall

> *Make no mistake, we will ask more and more from our reservists, from operational commitment to training support to close recce, to numerous specialties which they, and they alone will have.*[2]
>
> Gen. Hillier

Transformation is a word that has a unique connotation for members of the Canadian Forces (CF). For some, the word conjures emotions of fear and uncertainty, while for others it holds the promise of opportunity. Coined to describe the process by which the CF will adjust its approach to the new reality of full spectrum operations,[3] everything from command structures to the operational art – the strategy, operations, and tactical approach governing the way the CF fights – has and will come under scrutiny.

Among the many changes to be addressed is the need to more successfully integrate larger numbers of reserve force personnel (augmentees) into line battalions preparing for overseas deployment. Since the beginning of the 1990s, the marked increase in the number of peacekeeping operations has necessitated a shift in thinking within the CF and allowed ever-larger numbers of reservists to volunteer for overseas service. The result has been the opening of a Pandora's Box of administrative and personnel problems that needed to be addressed if integration was to be successful. While the ensuing fifteen years of high paced operations have created a large number of administrative work-arounds, the larger issue of how to successfully integrate the augmentees into the section/platoon has been largely left to the junior leadership, a practice that is not unique to the Canadian Forces. Indeed, Britain, Australia, New Zealand and the United States, each with varying degrees of success, have all had to address this same issue.

At the heart of the problem lies the desire by command to create unit integrity and the esprit de corps so important to survival when conducting full spectrum operations. To this end, it is well recognized that some of the factors that contribute to esprit de corps among military groups in peacetime are compatibility, conformity, comradeship, competition, tradition, territoriality and status. These and other factors are also vital in combat environments and are reinforced by recognition, combat and the importance of interdependence.[4] Yet, when two unknown groups come together, it is natural that "fear of the unknown" will cause the ranks to close against outsiders. Further, each soldier will bring their own form of "baggage" based on their own experience that colours their approach to the integration issue. Since all Canadian soldiers have been indoctrinated into the Regimental System, the attitudes and approaches fostered there can, and sometimes do, lead to an adversarial environment. The purpose of this paper is to discuss some of the problems faced when integrating reserve and regular force troops and to suggest some ways in which to reduce those problems.

HISTORY

For the Canadian military the issue of integration is not new. The issue was first highlighted by Sedentary Militia in the early 1800s. The performance of the militia during the War of 1812, coupled with an over zealous desire by some to claim credit for their accomplishments, led to the creation and later propagation of what has become know as the "Militia Myth."[5] This myth held that it was the militia – with the help of a few regular army troops – who had saved Canada during the war with the Americans.

History has shown the myth to be just that, a myth. That Canada survived at all during its early years is probably more attributable to good luck than the fighting ferocity of its Militia. In every war fought in North America until the American Civil War, the brunt of the fighting was borne by regular troops, French, British, or American. Regular officers planned the campaigns, husbanding their resources of trained men. Militias were useful, sometimes critically so, but Militiamen fresh from the farm or the shops of the towns would not and could not stand up to the rigors of the battlefield.[6] That is not to imply that the Militia did not acquit themselves well in battle. On the contrary, in October 1813, for instance, a mixed force of regulars and Militia numbering approximately 1700 men in various states of training, forced the retirement of a superior but ill-led American force of 3000-4000 after a short, sharp engagement that saved Montreal.[7]

As Canada gradually took control of its destiny in the late 1800s the creation of the first permanent force units to help the militia train brought further attention to the issue. The militia considered the early permanent force units to be nothing more than the dregs of society. Conversely, the permanent force and professional soldiers in command positions often lamented the club-like atmosphere of the militia unit. The war of words that was waged between the two groups was frequently vitriolic and protracted, as each side remained firmly entrenched and unswayed from its belief that it was the pre-eminent force.

Much of the problem could be said to be centred on the discrepancy in the level of training between the two forces. It can also be said that in the early days this problem was largely overcome by time spent in the field working together. However, as the level of sophistication in warfare increased, the time required for training augmentees up to the required standard correspondingly increased which in turn only served to highlight the discrepancy.

SHIFTING EMPHASIS

From the Boer War to the end of the Korean War, the regular and reserve forces continued to jockey for position as Canada's pre-eminent defence force. The wars of the late nineteenth and first half of the twentieth century had largely levelled the playing field as far as skills within the combat arms were concerned; yet skill fade within the militia ranks was a real issue between the wars.

By the late 1950s, the advent of tactically deployable nuclear weapons forced Canada to re-examine its requirement of the military. The result was a shift in emphasis from a force generation capability, to a highly trained rapidly deployable force modeled on the Korean experience. Chief of the General Staff (CGS) Guy Simmons determined that the Militia was not capable of providing trained men immediately available for rapid mobilization to meet a sudden crisis.[8] Once subordinate to the Militia, the regular force now surged ahead taking the leading role in the defence of Canada and leaving the militia with few opportunities for "real soldiering." The ensuing gradual, but inevitable, decline in the standard of training of the militia soldier led to the emergence within the regular force of the belief that the militia were nothing more than weekend warriors.

For the thirty years following Korea, the regular force occupied the place of prominence while the militia languished on the sidelines with no clearly defined role or opportunity to perform the tasks for which they trained. In 1987, the federal government attempted to address the situation by releasing a White Paper that embraced the concept of a Total Force. It spoke about one army, and embarked on an experiment in which some battalions and regiments would be regular/reserve amalgams, and would have specific, order-of-battle related roles in a rejuvenated CF. Thus was born the 10/90 battalion. Indeed, the promise implicit in the proposal was soon realized when reservists found themselves on the front-line at Oka and actively supporting Operation Friction. But thirty years of high-readiness training within the regular force again only highlighted the discrepancy between regular and reserve soldiers, and few remained convinced of the concept of integration.

TROUBLED RELATIONS

The demise of the Iron Curtain, however, again shifted the operational pace for the CF into high gear with a quick succession of deployments to Yugoslavia, Rwanda, Cambodia and Somalia. Faced with drastically under strength battalion establishments caused by long-term government policies of force reduction, the regular force, out of necessity, turned to the militia to fill in the ranks. So much so that on some of the early deployments to Yugoslavia as many as 50 per cent of some of the companies consisted of reservists.[9]

At the command level, however, despite a sudden increase in the number of reserve augmentees, by 1995 the relationship between the regular and reserve command was not on a good footing. The regular force, given the increased operational pace, felt the militia should do as they were told and continue providing a ready source of augmentees. Their focus was on meeting the ever-increasing operational needs and not on regimental survival. To this end, and in an effort to meet their force-generation targets, the regular force offered direct entry to as many reserve soldiers as would come.

The reserve establishment, on the other hand, distrustful of the regular forces' intentions and seeing their ranks being depleted faster than they could recruit, fought to oppose any change to its traditional role of national mobilization and to keep all of its regiments – no matter how small – alive. Each organization's distrust of the other grew to such a state that by 1997 they had reached the "low water" mark in the history of their relations.[10] And thus the lines were drawn.

A QUESTION OF LEADERSHIP

The 1990s represented a period in which the operational pace for the CF increased many times over. In the former Yugoslavia, for example, there were no fewer than 62 separate and distinct operations mounted between 1991 and 2003.[11] Throughout, the regular and reserve forces had to learn to work together more effectively. Administrative work-arounds dealing with the issues of pay, contracting and benefits packages were being addressed but had (and have) yet to be consolidated resulting in some reserve soldiers being treated differently than their regular force counterparts.[12] An example of this comes from early in the ramp up training for Operation Harmony, Rotation (Roto) 5.

A reservist's wife was reportedly admitted to emergency in Ottawa while he was in training in Petawawa. Told to report to the Officer Commanding (OC), the soldier was informed that he would not be permitted to leave training and go to Ottawa. The reason given was that since he did not know if the individual was entitled to emergency leave under his Class C contract, the OC wanted to save the soldier's leave for a *real* emergency. Though his wife was fine in the end, the episode aptly demonstrates the OC's lack of understanding of the nuances of the terms and conditions of service of men under his command.

Other issues also complicate the integration of augmentees into regular battalions. Junior leaders often fostered the "us versus them" attitude at the platoon level. Again, early on in the training for Operation Harmony, Roto 5, the initial selection of the augmentees had not yet begun. Many of the reservists were still trying to figure out where they were in the pecking order and what was expected of them. During a battalion parade one of the reservists had the misfortune of collapsing due to heat exhaustion and had to be helped off the parade square. This then led to a tirade by the platoon warrant officer. Talking to "his boys" (the regulars force troops in his platoon), he began to run down the "weekend warriors" describing in great detail – but quietly enough so that only his platoon could hear him – how he was going to run the reserves in to the ground and how they were not worth wasting his time on. This had the intended detrimental effect upon the reservists and their morale promptly plummeted.[13]

Arguably, both examples cited illustrate a lack of leadership ability on the part of the individual leaders involved. Indeed, following the ten leadership principles espoused by the CF, the platoon warrant officer failed to "know his soldiers and promote their welfare."[14] And, the company commander did not make a "sound and timely decision."[15] Would each of the leaders have treated a regular force member the same? Or do these examples speak to the level of institutional discrimination inherent in the Army at the time? Indeed, this discrimination had degenerated into name-calling on many occasions and can still be heard today.[16] Regardless, the 1990s represented the first real large-scale exposure the regular force had to large numbers of reservists in their ranks and thereby underscored some of the problems of integration.

A NEW RELATIONSHIP

The times, however, were changing. Despite the "weekend warrior" reputation, reservists volunteering for overseas service brought more than basic soldiering skills to the table. Many had educations that were better than their regular force counterparts and some brought skills that were not readily available in the combat arms trades.[17] Long exposed to poorly trained militiamen, both of these factors served to open the eyes of some in the regular force and earned the militia more respect. Of course, the fact that the bulk of reservists acquitted themselves exceptionally well while on tour has served to impress upon the regular force that the reserves bring one critical factor to the table over and above just their basic skills: they bring their desire to be there. The ensuing fifteen years of exposure to well trained and disciplined reservists have done much to dispel the notion of "weekend warrior" and garner the augmentee greater respect within the regular force ranks.

The expanded role planned for the militia in future operations as articulated by the Minister of National Defence in the Defence Policy Statement,[18] forecasts a much closer working relationship between the regular and reserve forces and sets the stage for a revised working relationship. While it is true that there continues to be a discrepancy between the training and experience levels of the two forces, great strides have been made to reduce the gap (at least on the training side) by Land Force Reserve Restructure (LFRR) with the adoption of the One Army, One Standard policy.[19] This standardization in training has been bolstered by a more recent phenomenon. Not only are reserve soldiers returning from tour with valuable experience to pass on, but across Ontario reserve units are seeing increased numbers of regular members with twenty or more years of experience leaving the regular force only to join the local reserve unit so that they can continue on with the soldiering that they so love. The result has been a gradual leveling of the playing field between the two forces. The evidence of this new direction can be seen in Task Force 03/06 to Afghanistan in which 220 reserves will deploy as part of the main contingent at ranks from private to major, proof that the reserves are being recognized at their rank and experience level as equals to their regular force counterparts. Today, it is expected that reserve soldiers of all levels will have received the same training as their regular force counterparts with the only discriminating factor being the amount of experience the soldier brings to the table.

Still, any time outsiders join an already formed body, be it troops in battalion or professionals in an office setting, there will be a feeling out period in which the group tests the new individual(s) to determine whether they will be accepted. In his book *Men Against Fire*, S.L.A. Marshall points out that just as men will not fight for a leader they do not know, the soldier has an inherent unwillingness to risk danger on behalf of men with whom he has no social identity.[20] This is particularly true regarding new augmentees. With this in mind, the question of how does the leader/commander integrate new augmentees into his command surfaces. And, what does the augmentee have to do to prove that he/she belongs? To answer this question, one must first have an understanding of small group dynamics.

In an article titled "Developmental Sequence in Small Groups," Bruce Tuckman outlines the five stages of group development as being: Forming; Storming; Norming; Performing; and Adjourning.[21]

In Stage one – forming – group members rely on safe, patterned behaviour and look to the group leader (in our case, the section commander) for guidance and direction. They have a desire for acceptance by the group and a need to know that the group is safe. They set about gathering impressions and data about the similarities and differences among them and forming preferences for future sub-grouping.[22] In the military context, in order for the group to feel safe the existing group must feel comfortable with the incoming troops. And, in order for that to happen, the new troops must convince the group they are joining that they are capable and worthy of respect.

For the augmentee respect is garnered in a number of ways. First and foremost, it is to be remembered that the Army, and in particular the combat arms trades, are a profession largely based on one's *physical fitness*. The reservist must arrive at battalion ready to meet – and/or exceed – the basic physical training (PT) level. Being capable of strapping on a 55 pound backpack, helmet, webbing, and rifle, and marching 10 miles in the allotted time begins to earn the individual a measure of respect.

Next, *technical competence* plays a large factor in acceptance within an infantry section. Being able to take a C6 medium machinegun (or any other of the platoon weapons) and strip, clean, reassemble and perform the function test without asking for help means, in the eyes of those watching, that the individual does not require additional training or supervision and thus builds trust in the individual's competence.

Being *situationally aware*, also plays a large part in acceptance as it speaks further to the individual's level of training. Knowing where to go and what to do without being asked, or told, helps to instill confidence amongst the members and helps to build the team. It is worth noting that these three points will require much of the reserve soldier. In a three-hour-per-week army, it is implicit that the individual volunteering for overseas service will need to devote personal time to his training over and above his regular obligations in order to meet or exceed the acceptable standard.

Stage two of group dynamics – storming – which deals largely with attitude and personality, is likely the most important determining factor on whether the individual is accepted by the group or not. As the group members organize for the upcoming task, conflict inevitably results in their personal relations. Individuals have to bend and mould their feelings, ideas, attitudes and beliefs to suit the group organization. Because of "fear of exposure" or "fear of failure" there will be an increased desire for structural clarification and commitment.[23] It is here that the junior leader (section second in command (2IC), Section Commander, Platoon 2IC, and Platoon Commander) becomes all-important. Solid leadership at this level, by knowing your men and employing them to the level of their ability,[24] can make or break an organization. An individual's ability to "blend" with the existing group can determine whether or not someone is accepted into the group. Nothing spells rejection faster than an individual with the attitude that they know it all and are unwilling to learn.

However, acceptance is a two way street and the strongest builder of bonds within a section or platoon, is shared adversity. An event, or task, that forces the group to work together as a team to accomplish the mission helps foster acceptance. This can be something as simple as a particularly gruelling PT run, or as harrowing as a firefight. In the end, the team will begin to gel because of the realization of a shared need to rely upon each other and the testing of that need under trying conditions. Regardless of each individual's background, the trust built under adverse conditions can cement friendships and build effective working units.

For the regular force junior leaders charged with making the sections and platoons work effectively, the adherence to, and application of, the principles of leadership, combined with the need to recognize that the individual reservist who appears on the parade square ready to train does so at personal sacrifice, helps make the integration effective. In many cases, the reservist has either quit a job or put his education on hold in order to avail himself of the opportunity to serve his country and go overseas. Moreover, that individual may well have skills that are not readily available within a combat arms trade but that may come in handy in situations outside the normal job description of the soldier while deployed. This demonstrated desire to put their life "on hold" is worthy of respect.

CONCLUSION

The desire to serve one's country is a long-standing tradition in Canada. From the late 1600s right up to the present day, many are the examples of Canadians setting aside their tools to take up arms to defend our nation. But the world is a changing place, and to meet the challenge of the increased tempo of full spectrum operations, our armed forces are undergoing a major transition. The endstate will be nothing short of a full working partnership between the regular and reserve forces.

For the reservist, it is no longer sufficient to pick up a rifle and stand shoulder to shoulder with his regular force counterpart to ward off the enemy. Today much more training is required. The reservist intent on deploying overseas must arrive at battalion physically fit, technically proficient, situationally aware and prepared to begin training. To accomplish this requires a devotion of personal time over and above regular training. In short, he must become a full-time soldier on a part-time basis.

The regular force too, at the junior leader level, must accept that it can no longer meet the increased operational tempo without the help of the reservist as it integrates militia augmentees into the ranks. This is best achieved by the application of, and adherence to, the leadership principles. Furthermore, regulars must come to recognize and accept that skill-sets not traditionally found within the ranks are strengths rather than weaknesses. Finally, regular force members must recognize that reservists bring with them an intense desire to be there, often times at personal sacrifice.

The ghosts of past reputations will continue to haunt the CF as long as individuals and institutions hold on to preconceived notions about each other. In an environment as unstable as full spectrum operations, animosities between individuals cannot be afforded, since any doubt about levels of training or ability to execute the job could well lead to the loss of life. Much work is required by both sides to ensure that appropriate trust is built and fostered within the section and platoon. Certainly, there can be no more appropriate place to build this trust than through the shared adversity of pre-deployment training.

ENDNOTES

1 S.L.A. Marshall, *Men against Fire* (New York: William Morrow and Company, 1947), 115.

2 Lt. Gen Hillier, *Army Transformation, Punching Above our Weight.*

3 Full Spectrum Operations is the CF term used to describe the Three Block War which is a concept devised by Marine General Charles Krulak in the late 1990s to describe the complex spectrum of challenges likely to be faced by soliders on the modern battlefield. In three contiguous city blocks soldiers may be required to conduct full scale military action, peacekeeping operations and humanitarian relief.

4 Anthony Kellett, *Combat Motivation, the Behavior of Soldiers in Battle* (Boston: Kluwer-Nijhoff, 1982), 97.

5 Jack Granatstein, *Canada's Army* (Toronto, 2002), 10.

6 Ibid.

7 Ibid., 11

8 The following discussion is based largely on the report titled *Militia Reform* by Dr. Steven Harris, senior historian with Department of History and Heritage (DHH), written on 30 March 1995 and updated 13 January 2002.

9 Personal interview with CWO J.O.M.P. Leblanc, Royal 22nd Regiment, B Company Sergeant-Major on Operation Cavalier.

10 Interview with Major-General Fitch, October 2005.

11 This figure is based on the research conducted by the author as part of his duties with the Department of National Defence's DHH War Diaries team.

12 Interview with Commander E. Naismith, COS DGMHRPP, ADM (HR-Mil), December 2005.

13 Both examples cited are based upon the authors own experience and are recounted in the book *All Tigers No Donkeys, A Canadian Soldier in Croatia* (St. Catherine's, 2004).

14 From the *Principles of Leadership* poster published by the CF, 2004 edition, Leadership principle #5.

15 Ibid., Leadership principle #8

16 The regular force often referred to the reserves as "toons." Short for cartoons, it implied that the reserves wear colourful hats; were hugely entertaining; and came out only on weekends. The reserves on the other hand, referred to the regular force as "soaps,' implying that they were enormously boring; hugely melodramatic; and were done by 2:00 on a weekday afternoon.

17 The following is an example of the skills brought by 3 reservists to one section alone: Carpenter, Painter, Commercial helicopter pilot, industrial engineering technician, three university degrees, two technical certificates, and 7 military trade qualifications .

18 Minister of National Defence, *Defence Policy Statement, A Role of Pride and Influence in the World*, June 2005.

19 Capt. M.T. Aucoin, *The Army Doctrine and Training Bulletin, One Army*, One Standard; Vol. 6 No. 3 Fall/Winter 2003.

20 Marshall, *Men against Fire*, 153.

21 B. Tuckman, "Developmental Sequence in Small Groups," *Psychological Bulletin*, 63 (1965), 384-399. In the article, Tuckman outlines five stages of group development. **Stage 1** is *Forming*. In the forming stage, personal relations are characterized by dependence. Group members rely on safe, patterned behaviour and look to the group leader for guidance and direction. Group members have a desire for acceptance by the group and a need to know that the group is safe. They set about gathering impressions and data about the similarities and differences among them and forming preferences for future subgrouping. In **Stage 2**, *Storming*, is characterized by competition and conflict in the personal-relations dimensions and organization in the task-functions dimension. As the group members attempt to organize for the task, conflict inevitably results in their personal relations. Individuals have to bend and mould their feelings, ideas, attitudes, and beliefs to suit the group organization. Because of "fear of exposure" or "fear of failure" there will be an increased desire for structural clarification and commitment. Although conflicts may or may not surface as group issues, they do exist. **Stage 3** is labelled *Norming*. In this stage, interpersonal relations are characterized by cohesion. Group members are engaged in active acknowledgement of all member contributions, community building and maintenance, and solving of group issues. Members are willing to change their preconceived ideas or opinions on the basis of facts presented by other members, and they actively ask questions of one another. Leadership is shared, and cliques dissolve. When members begin to know – and identify with – one another, the level of trust in their personal relations contributes to the development of group cohesion. **Stage 4** is labelled *performing*, and is not always achieved by the group. In this stage, people can work independently, in

sub-groups, or as a total unit with equal facility. Their roles and authorities dynamically adjust to the changing needs of the group and individuals. **Stage 5**, *adjourning*, involves the termination of task behaviours and disengagement from relations. A planned conclusion – such as the end of a tour – usually includes recognition for participation and achievement and an opportunity for members to say personal goodbyes.

22 Ibid, stage 1.

23 ibid, stage 2.

24 The ten principles of leadership taught within the CF are as follows **1)** Achieve professional competence, **2)** appreciate your own strengths and limitations and pursue self-improvement, **3)** Seek and accept responsibility, **4)** lead by example, **5)** Know your soldiers and promote their welfare, **6)** Make sure that your followers know your meaning and intent, then lead them to the accomplishment of the mission, **7)** Develop leadership potential in your followers, **8)** Make sound and timely decisions, **9)** Train your soldiers as a team and employ them up to their capability, and **10)** Keep your followers informed of the mission, the changing situation, and the overall picture.

CONTRIBUTORS

Chief Warrant Officer **Alex Bouzane** enrolled in the Canadian Forces on 18 April 1969. He has served with 414 Sqn, North Bay, 1 CAD Germany, 103 RU, Gander, 431 Sqn (Snowbirds) and BAME Organization Moose Jaw, 450 Sqn Ottawa, 419 and 416 Sqn Cold Lake. He was Wing Chief Warrant Officer of Goose Bay, Task Force CWO in Aviano and Detachment CWO TAL Det, SWA.

Chief Petty Officer First Class **Eric M. Carberry** enrolled in the Canadian Forces in January 1976 as a Radar Plotter. He served in this capacity until 1985 when he became a Naval Electronics Technician. CPO1 Carberry has served on HMCS *Annapolis*, *Nipigon*, *Gatineau* and *Iroquois*. His shore postings have included the Naval Electronics School, MARCOM Headquarters, CFB Shearwater and Halifax where he served as the Trials Officer Fleet Maintenance Facility Cape Scott and as Unit Chief Petty Officer for Fleet Maintenance Facility Cape Scott. He was appointed Base Chief Petty Officer, CFB Halifax, in June 2005 and the Maritime Forces Atlantic Chief Petty Officer in 2006.

Chief Warrant Officer **Kit Charlebois** was the Task Force Sergeant-Major for Operation Athena, Rotation 3.

Chief Warrant Officer **J.A. Comeau** enrolled in the Canadian Forces in February 1971. His experience in the CF includes three years of service with the Reserves and 31 years of service with the Regular Force.

Chief Petty Officer First Class (ret'd) **Pat Evans** joined the Navy as a Radio Operator in January 1972. Over the course of his career, he became a Radio Technician and eventually served two years as the Career Manager for the Naval Electronic Technicians. Upon completion of his tour as coxswain on HMCS *Iroquois*, CPO1 Evans was appointed Chief of the Canadian Forces Naval Engineering School at CFB Halifax where he retired in July 2006.

Sergeant **Kurt Grant** is a published author and has written several articles on military and militia matters. He is currently employed at the Department of National Defence Directorate of History and Heritage in Ottawa. He is also a serving member of the Brockville Rifles.

Colonel, Dr. **Bernd Horn** is an experienced infantry officer and professor. He has command experience at the sub-unit and unit levels and was the officer commanding 3 Commando, the Canadian Airborne Regiment, from 1993-1995 and the commanding officer of the 1st Battalion, The Royal Canadian Regiment from 2001-2003. He is also an adjunct professor of history at the Royal Military College of Canada. Colonel Horn is currently serving as the Deputy Commander of Special Operations Forces Command

Chief Warrant Officer **André Jutras** served as the Regimental Sergeant-Major (RSM) of 5 Canadian Service Battalion during deployment to Afghanistan on Operation Athena, Rotation Zero, from February to August 2004. During this deployment, he was the RSM for the National Support Element (NSE).

Chief Warrant Officer **Charles Malchow** joined the CF in November 1976. As a Construction Engineering Technician (drafting & survey), he has been posted to CFB Gagetown, CFB Ottawa, CFS Beaverlodge, CFB Winnipeg, CFB Chilliwack, 4 Wing Cold Lake, 1 CEU Moncton and he is presently employed as the Contracts Officer and Unit Chief Warrant Officer with the Engineer Services Company at CFB/ASU Edmonton. His operational deployments include: Golan Heights, CFS Alert, Bosnia, Aviano Italy and South West Asia.

Mr. Craig **Leslie Mantle** graduated from Queen's University in 2002 with an MA in Canadian military history and he is currently employed as a researcher at the Canadian Forces Leadership Institute.

Master Warrant Officer **T.J.P. Mugford**, formerly of the Royal Canadian Regiment, served as the Camp Sergeant-Major, Kabul, Afghanistan during Operation Athena, Rotation Zero.

Chief Petty Officer First Class **Ken Mullins** joined the Royal Canadian Navy in 1977 and has served on numerous ships as a Weaponsman Surface and Naval Weapons Technician. He was promoted to Chief Petty Officer First Class in December 2000 and was appointed Coxswain of HMCS *Montreal* from May 2002 until August 2004. He is currently serving as the Ammunition Inspector for Maritime Forces Atlantic in Halifax.

Chief Warrant Officer **Randy Northrup** joined the Canadian Forces in August 1977. He has served in all three battalions of the PPCLI, as well as with the Canadian Airborne Regiment. He served as a CSM in Croatia, from April to October 1993, and as 1 PPCLI Battle Group RSM in Kandahar, Afghanistan from January to August 2006. CWO Northrup received the Meritorious Service Cross for his service in Afghanistan. He is currently the RSM of the Canadian Manoeuvre Training Centre, CFB/ASU Wainwright.

Chief Warrant Officer **Michael Poirier** served as the Regimental Sergeant-Major on Operation Athena, Rotation 1.

Chief Warrant Officer **Thomas J. Secretan** has over 12 years experience in rank and has been a member for the Canadian Forces for over 33 years. He has held the position of Squadron Chief Warrant Officer in three different Squadrons and has been the Chief Warrant Officer for two overseas deployments and one domestic operation. He currently holds the position of Wing Chief Warrant Officer, 8 Wing, Trenton, Ontario.

Dr. **Emily Spencer** holds an MA and PhD in War Studies from the Royal Military College of Canada. She has taught Canadian and Canadian military history at the Royal Military College of Canada. She is currently employed as a Research Associate by the Canadian Forces Leadership Institute.

Chief Warrant Officer **Lee Topp** joined the Canadian Forces in 1974 as a crewman in the Armoured Corps. He served his first 20 years on regimental duty with the 8th Hussars in Petawawa, as well completing tours in Cyprus and Germany. He has also been employed at the Canadian Forces Officer Candidate School in Chilliwack, the Combat Training Centre (CTC) in Gagetown, the Land Forces Trials and Evaluation Unit (LFTEU) and the Army Individual Training Development in the Combat Training Centre Headquarters. He served in Rwanda as a CSM in 1995 and as a RSM on Operation Athena, Rotation 3, in Afghanistan in 2005. Upon his return, he was appointed to CTC as the formation sergeant-major. In 2007, Captain Topp took his commission and returned to LFTEU.

GLOSSARY
OF ACRONYMS AND ABBREVIATIONS

2IC	Second-in-Command
4GW	Fourth Generation Warfare
AAR	After Action Review
AJAG	Assistant Judge Advocate General
ANA	Afghan National Army
ANP	Afghan National Police
ANSF	Afghan National Security Forces
AO	Area of Operations
AOR	Area of Responsibility
APRV	Annual Personnel Readiness Verification
ARG	Amphibious Ready Group
ASIC	All Source Intelligence Centre
ASF	Airfield Security Flight
ATA	Afghan Transitional Authority
BCT	Brigade Combat Team
BG	Battle Group
Bn	Battalion
CANCAP	Canadian Forces Contractor Augmentation Program
CAS	Chief of the Air Staff
CBC	Canadian Broadcasting Corporation
CCPO	Command Chief Petty Officer
CDS	Chief of Defence Staff
CEU	Construction Engineer Unit
CF	Canadian Forces
CFB	Canadian Forces Base
CFPSA	Canadian Forces Personnel Support Agency
CFHQ	Canadian Forces Headquarters
CFLCC	Canadian Forces Land Component Command
CGS	Chief of the General Staff
CIDA	Canadian International Development Agency
CIMIC	Civil-Military Cooperation
CLP	Combat Logistics Patrol
CMBG	Canadian Mechanized Brigade Group

CMS	Chief of Maritime Staff
CO	Commanding Officer
CoC	Chain of Command
Col	Colonel
Coy	Company
CP	Command Post
Cpl	Corporal
CPO	Chief Petty Officer
CPO1	Chief Petty Officer First Class
CPO2	Chief Petty Officer Second Class
CQ	Company Quartermaster
CSM	Company Sergeant-Major
CSN	Control System Network
CTC	Combat Training Centre
CTF	Canadian Task Force
CWO	Chief Warrant Officer
DAG	Deployment Assistance Group
DCDS	Deputy Chief of the Defence Staff
DCO	Deputy Commanding Officer
DDIO	DCDS Directives for International Operations
DFA	Department of Foreign Affairs
DIN	Defence Information Network
DND	Department of National Defence
DWAN	Defence Wide Area Network
EOD	Explosive Ordnance Disposal
ETT	Embedded Training Teams
EW	Electronic Warfare
FARP	Forward Armament and Refuel Point
FOB	Forward Operation Base
FOD	Foreign Object Damage
FST	Forward Surgical Hospital
GPS	Global Positioning System
HLTA	Home Leave Travel Allowance
HQ	Headquarters
HMCS	Her Majesty's Canadian Ship

IC	In Command
ID	Identification
IED	Improvised Explosive Devise
IRF (L)	Immediate Reaction Force (Land)
ISAF	International Security Assistance Force
ISO	International Standards Organization
JTF2	Joint Task Force 2
KAF	Kandahar Airfield
KMBG	Kabul Multi-National Brigade Group
KMNB	Kabul Multinational Brigade
KRWAU	Kosovo Rotary Wing Aviation Unit
LAN	Local Area Network
LAV	Light Armoured Vehicle
LAVW	Light Armoured Vehicle Wheeled
LDSH (RC)	Lord Strathcona's Horse (Royal Canadians)
LFAA	Land Force Atlantic Area
LFCA	Land Force Central Area
LFRR	Land Forces Reserve Restructure
LFTEU	Land Forces Trials and Evaluation Unit
Log O	Logistics Officer
LRP	Long Range Patrol
LRPTF	Long Range Patrol Task Force
LSVW	Light Support Vehicle Wheeled
LTA	Leave Travel Allowance
LUVW	Light Utility Vehicle Wheeled
LZ	Landing Zone
MCpl	Master Corporal
MEL	Master Equipment List
MFRC	Military Family Resource Centre
MILES	Multiple Integrated Laser Engagement System
MNB	Multi-National Brigade
MND	Minister of National Defence
MP	Military Police
MSE	Mobile Support Equipment
MWO	Master Warrant Officer

NATO	North Atlantic Treaty Organization
NCCIS	National Command, Control and Information System
NCE	National Command Element
NCM	Non-Commissioned Member
NCO	Non-Commissioned Officer
ND	Negligent Discharge
NDHQ	National Defence Headquarters
NGO	Non-Governmental Organization
NSE	National Support Element
NSU	National Support Unit
NVG	Night Vision Goggles
OC	Officer Commanding
O-group	Orders Group
Op	Operation
OPCON	Operational Control
OR	Operating Room
ORBATS	Orders of Battle
PAO	Public Affairs Officer
PD	Professional Development/Police District
PER	Personnel Evaluation Reports
PO	Petty Officer
POL	Petroleum, Oil and Lubricants
PPCLI	Princess Patricia's Light Infantry
PRT	Provincial Reconstruction Team
PT	Physical Training
PVC	Polyvinyl Plastic
PWGSC	Public Works Government Service Canada
QRF	Quick Reaction Force
R22eR	Royal 22nd Regiment
R&R	Rest and Relaxation
RC	Roman Catholic
RCD	Royal Canadian Dragoons
RCHA	Royal Canadian Horse Artillery
RCR	Royal Canadian Regiment
Recce	Reconnaissance
Ret'd	Retired

RMS	Resource Management Support
RoE	Rules of Engagement
RON	Remaining Over Night
Roto	Rotation
RPG	Rocket Propelled Grenade
RSM	Regimental Sergeant-Major
RTC	Relocatable Temporary Camp
RTU	Return to Unit
SAT	Site Activation Team
SAV	Staff Assistance Visit
SET	Special Engineering Team
SF	Special Force
Sgt	Sergeant
SITREP	Situation Report
SIV	Staff Inspection Visits
SM	Sergeant-Major
SOFA	Standing of Force Agreement
SOPs	Standard Operating Procedures
Sqn	Squadron
Sr	Senior
SWA	South West Asia
Tac	Tactical
TAL Det	Tactical Airlift Detachment
TAT	Theatre Activation Team
TAV	Technical Assistance Visit
TD	Temporary Duty
TF	Task Force
TFK	Task Force Kabul
TMST	Theatre Mission Specific Training
TOCA	Transfer of Command Authority
TOETS	Test of Elementary Training
TOR	Terms of Reference
Tpt	Transport
TTPs	Tactics, Techniques and Procedures
TRADOC	Training and Doctrine Command
TSE	Theatre Support Element

UAE	United Arab Emirates
UN	United Nations
US	United States of America
VBIED	Vehicle Borne Improvised Explosive Device
VIP	Very Important Person
VVIP	Very, Very Important Person
VOR	Vehicle on Repair
WainCon	Wainright Training Area Concentration
WO	Warrant Officer
XO	Executive Officer

INDEX

11 September 2001 (9/11) x, 1, 2, 13, 17, 21, 22, 27

Adriatic 66
Afghan 29, 79, 115, 117, 124, 125, 126, 127, 128, 134, 142, 146, 149, 150, 155, 156, 157, 158, **193** *gloss.*
Afghan National Army (ANA) 29, 117, 126, 127, 134, 146, 155, 156, **193** *gloss.*
Afghan National Police (ANP) 126, 127, **193** *gloss.*
Afghan National Security Force 127, **193** *gloss.*
Afghan Transitional Authority (ATA) 115, 179, **193** *gloss.*
Afghanistan xi, xv, xxii, 1, 5, 23, 24, 27, 28, 32, 34, 35, 83, 95, 96, 98, 107, 110, 112, 117-119, 121, 123-126, 128, 129, 131, 133, 135, 142, 145, 153, 158, 162, 167, 172, 181, 190, 192
Air Force 11, 28, 75, 83, 84, 92, 102, 142
Airfield Security Flight (ASF) 86, **193** *gloss.*
Al-Qaeda 15, 22, 30, 32, 34, 55, 69, 79
Alcohol 25, 70, 71, 94, 104
American Special Forces 34
Amphibious Ready Group (ARG) 15, 16, **193** *gloss.*
Area of Operation (AO) 69, 128, 132, 133, 139, 142, 143, 159, 166, 170, 171, **193** *gloss.*
Area of Responsibility (AOR) 15, 67, 95, 98, 105, 115, **193** *gloss.*
Army vi, ix-xxii, **xxiv-xxvii** *notes*, 11, 18, 24, 25, 29, 83, 84, 88, 103, 111, 117, 119, 122, 126, 127, 134, 142, 146, 155, 175, 176, 178, 180-182, **185** *notes*, **186** *notes*, 191, **193** *gloss.*
Augmentees 3, 85, 138, 159, 175-177, 179-181, 184
Aviano 39, 90, 189, 190

Bagram 28-31, 33, 141, 157
Banyans 70
Bin Laden, Osama 21, 33
Boarding Parties 57
Bosnia-Herzegovina 119
British **xxv** *notes*, **xxvi** *notes*, 24, 84, 89, 92, 125, 157, 162, 177
British Army **xxvi** *notes*, 24, 84

Cahar-Asyab 134
Cambodia 178
Camp Commander 41, 80, 96
Camp Defence 80
Camp Julien 115, 116, 142-144, 148, 152
Camp Mirage 1, 5, 10, 47, 83, 95-105, 141, 142, 148, 153
Camp Wainwright 86
Canada iv, xiii, xv, **xvii** *notes*, **xxiv** *notes*, **xxv-xxvii** *notes*, 3, 6, 7, 11, 13, 14, 16, 17, 25, 28, 31, 33-35, 41, 50, 51, 55-60, 65, 66, 71, 73, 89, 90, 92-97, 100, 107, 112, 115, 122, 123, 128, 129, 135, 153, 154, 159-161, 165, 170, 173, 176-178, 184, **185** *notes*, 190, 191, **196** *gloss.*
Canadian Airborne Regiment xvii, 190, 191
Canadian Embassy 11, 142
Canadian Forces (CF) iii-vi, xv, **xxv** *notes*, **xxvii** *notes*, 1, 10, 13, 21, 39, 50, 66, 85, 97, 107, 112, 134, 163, 175, 176, 189-191, **193** *gloss.*
Canadian Forces Base (CFB) 1-3, 5, 36, 189, 190, **193** *gloss.*

Canadian Forces Personnel support Agency (CFPSA) 50, 97, 102, 103, 107, 134, 143, 146, 148, **193** *gloss.*
Canadian International Development Agency (CIDA) 142, 150, **193** *gloss.*
Canadian Mechanized Brigade Group 131, **193** *gloss.*
Canadian public vi, 25, 29
Casualty(ies) xii, xvii, xix, 31, 32, 67, 110, 126, 147, 156, 161
Change of Command 141, 148, 159
Chief of Defence Staff (CDS) 1, 2, 136, 154, 158, **193** *gloss.*
Chief of the Maritime Staff 60, **194** *gloss.*
Chief of the Air Staff (CAS) 90, 95, **193** *gloss.*
Chief Warrant Officer xv, **xxv** *notes*, 39, 83, 133, 189, 190, 191, **194** *gloss.*
Clancy, Tom xvi, **xxv** *notes*
Climate 44, 48, 68, 123
Coalition Units 100
Code of Service Discipline xiii, 138
Cold War vi, ix, x, xxii
Combat xi, xiii, xiv-xxi, **xxv** *notes*, **xxvi** *notes*, **19** *notes*, 22, 24, 30, 32, 34, 43, **64** *notes*, 90, 99, 121, 123, 124, 165, 170, 176, 177, 180, 182, 183, **185** *notes*, 191, **193** *gloss.*
Command Structure 4, 22, 41, 81, 83, 96, 135, 175
Command Team 15, 17, 18, **19** *notes*, *20* **notes**, 26, 58, 65, 66, 70, 73, 131, 137, 141, 148
Commander xvi, xix, xxii, 22, 28, 32-35, 39, 41, **63** *notes*, 69, 80, 96, 103, 117, 131, 132, 134-138, 140, 141, 145, 146, 148, 154, 159, 168, 169, 171-175, 180, 181, **185** *notes*, 190
Commander's Conference 147, 148
Commanding Officer (CO) 2, 15, 18, 22, 23, 27, 29-32, 34-36, 39-41, 43, 45, 53, 58, 59, 61, 63, 65, 66, 68, 81, 84-89, 93, 94, 97, 101-103, 107-109, 117, 118, 122, 123, 131, 132, 134, 137, 148, 190, **194** *gloss.*
Conventional Forces 21
Cultural barriers 10, 76
Customs 3, 51, 80, 116, 118
Cyprus 121, 191

Dahrlaman 143
de Reil, Pierre xxii
Decompression Period 160
Defence iii, iv, vi, x, xi, xix, xxii, **xxvii** *notes*, 1, 2, 8, 21, 24, 27, 30, 34, 52, 70, 79, 80, 85, 86, 96, 127, 133, 136, 154, 157, 158, 177, 178, 181, **185** *notes*, **186** *notes*, 189, **193-196** *gloss.*
Defence Information Network (DIN) 85, 92, **194** *gloss.*
Defence Wide Area Network (DWAN) 52, **194** *gloss.*
Delmas, Phillippe x, **xxiv** *notes*
Department of Foreign Affairs (DFA) 142, **194** *gloss.*
Department of National Defence (DND) iv, xxii, **185** *notes*, 189, **194** *gloss.*
Deputy Chief of Defence Staff (DCDS) 1, 136, 154, 158, 159, **194** *gloss.*
Deputy Commanding Officer 40, 101, 103, **194** *gloss.*
Deputy Detachment Commanding Officer 39
Detachment Commander 39, **64** *notes*, **76** *notes*
Development iii, v, vi, ix, xxi, 5, 127-129, 142, 144, 162, 182, **186** *notes*, 191, **193** *gloss*, **196** *gloss.*
Diego Garcia 34, 35
Diplomacy 117, 127

Discipline ix, x, xiii, xvi, xvii, **xxvii** *notes*, 46, 68, 72, 94, 95, 99, 106, 107, 110, 116, 138, 160, 171, 175, 181
Drug Lords 79
Dubai 17, 30, 31, 70, 154, 160

East Timor 119
Education xv, xxi-xxiii, **xxvii** *notes*, 10, 144, 161, 162, 180, 183
Eggleton, Art xxii, xxvii
Elite 69
Equipment v, xviii, xx, 2-4, 6, **19** *notes*, 23, 33-35, 55, 69, 82, 89, 92, 98, 107, 108, 122-125, 127, 134, 139, 152, 158, 171, **195** *gloss.*
Esprit de corps 99, 176
Ethics xxii, **xxvii** *notes*, 104
Europe x, xvi, **xxvi** *notes*, 39, 44, 51
Executive Officer (XO) 15, 56-58, 62, 63, 65, 68, **77** *notes*
Explosive Ordnance Disposal (EOD) 33, 134, **194** *gloss.*

Fatigue 48, 91, 101, 160, 172, 173
Fear xii, xxi, **xxvi** *notes*, 15, 48, 90, 91, 109, 126, 137, 144, 175, 176, 183, **186** *notes*
Fighter pilots 43
Fitness 49, 92, 102, 103, 182
Flight Commanders 84, 85
Force Multiplier 126
Force Protection 17, 47, 68, 80, 136, 141, 142, 144-148, 150, 167
Forward Operating Base (FOB) 1, 34, 166, 170, 171, **194** *gloss.*
Fourth Generation Warfare (4GW) xi, **193** *gloss.*
Franks, Frederick M. xvi, **xxv** *notes*
Friction xii, 41, 178
Friedman, George xii, **xxiv** *notes*
Full spectrum operations 175, 176, 184, **185** *notes*

Gender 10, 93
Genocide 87
Germany x, 7, 32, 143, 161, 189, 191
Government of Canada 13, 25
Governor General 49, 90
Guam 35, 36
Gulf of Oman 13, 55, 65
Gundy Ghar 170

Haiti x, xii, 119
Headquarters (HQ) xxiii, 5, 8, 9, 11, 22, 24, 34, 39, 52, 87, 88, 134, 144, 146, 148, 154, 156, 189, 191, **193** *gloss.*, **194** *gloss.*, **197** *gloss.*
Heart 30, 89, 149, 174, 176
Heroes 37
Hillier, Rick 158, 175, **185** *notes*
HMCS *Fredericton* 58, **63** *notes*, 74
HMCS *Montreal* 55, 56, 62, 63, 190
HMCS *Iroquois* 19, 58, **63** *notes*, 76, 189
Home Leave Travel Allowance (HLTA) 7, 46, 100, 101, 103, 104, 159, 160, 161, 171, 173, **194** *gloss.*
Honesty xiii, xv, 122
Hope, Lieutenant Colonel Ian 123, 126, 128
Host Nation 6, 8, 9, 51, 100, 104
Humanitarian 13, 44, 127, **185** *notes*
Hygiene 26, 80

Immediate Reaction Force Land (IRF(L)) 22, 118
Improvised Explosive Device (IED) xi, 124, 134, 147, 149, 155, 157, 169, **197** *gloss.*
International Security Assistance Force (ISAF) 24, 133, 142, 156, **195** *gloss.*

Iran 155
Iron Curtain 178
Italy 8, 9, 90, 190

Jalalabad road 155
Joint Task Force (JTF) 2 32, 39, 123, **195** *gloss.*
Joint Task Force South West Asia 39
Jowz Valley 146
Judge Advocate General 162

Kabul 34, 79, 83, 95, 96, 98, 99, 101, 103-105, 115-117, 132-134, 137-139, 141-143, 150, 153-157, 160, 190, **195** *gloss.*, **197** *gloss.*
Kabul Multinational Brigade (KMNB) 117, 133, 134, 146, 156. **195** *gloss.*
Kandahar Airfield 24, 26-30, 32, 34, 125, 167, **195** *gloss.*
Kandahar City 127
Korean War 28, 177
Kosovo 83, 84, 86-92, 94, **195** *gloss.*
Kosovo Rotary Wing Aviation Unit (KRWAU) 83, 85, 86, 88, 90, 93, **195** *gloss.*
Kowst 32
Kyber Pass 155

Leader xiii, xvi, xix, xxi, 87, 94, 106, 181-184, 186
Leadership challenges v, 17, 55, 56, 69, 75, 79, 83, 84, 106, 117
Lessons Learned iii, v, 11, 39, 79, 121, 122, 159, 162, 163
Logar 155, 157

Macedonia 85, 87, 89
Maillet, Paul xxii, **xxvii** *notes*
Marines xxi, 15, 27, 30
Marshall, S.L.A. 175, 181, 185, 186
McManus, John xvii, xviii, **xxv** *notes*, **xxvi** *notes*
Media ix, xxii, 28, 29, 59, 81, 167
Middle East 14
Military Police (MP) 17, 39, 46, 86, 105, 134, 150, 155, **195** *gloss.*
Militia xiv, **xxiv** *notes*, 176-181, 184, **185** *notes*, 189
Mine xx, 24, 26, 79. 82, 95, 98, 104, 133, 138, 146, 155-157, 169. 172, 173
Minister of National Defence xxii, **xxvii** *notes*, 181, **186** *notes*, **195** *gloss.*
Montgomery, Bernard Law xiii
Morale xiv, xviii, **xxvii** *notes*, 4-8, 13, 16-18, **19** *notes*, **20** *notes*, 26, 42-47, 49, 50, 52, 58-61, 63, 66, 69-73, 80, **82** *notes*, 89, 90, 93, 98-100, 103, 106-108, 116, 122, 126, 129, 171, 175, 180
Morton, Desmond ix, **xxiv** *notes*
Mujahadeen 24
Multicultural 82
Murray, Williamson xii, **xxiv** *notes*
Muslim 10, 31, 70, 93

National Command Element (NCE) 116, 117, 134-136, 146, 165, 168, **195** *notes*
National Command, Control and Information System (NCCIS) 41, **195** *gloss.*
National Defence Headquarters (NDHQ) 8, 34, **196** *gloss.*
National Support Element (NSE) 8, 81, 108, 116, 134, 190, **196** *gloss.*
Navy vi, 14, 17, 18, **19** *notes*, 60, 62, **63** *notes*, 65, 66, 75, **76** *notes*, 83, 103, 189, 190
Negligent Discharge (ND) 160, **196** *notes*
Non-Commissioned Member (NCM) **xxvii** *notes*, 31, 43, 44, 46, 47, 50, 51, 55, 68, 72, 75, 85, 122, 171, **196** *gloss.*

Non-Commissioned Officer (NCO) iii, v, vi, ix-xxiii, **xxiv** *notes*, **xxv** *notes*, **xxvii** *notes*, 28, 52, 85, 89, 94, 96, 97, 99, 100, 118, 119, 121-123, 129, 138, 143, 144, 154, 160, 162, 166, 173, **196** *gloss.*
Non-Governmental Organization (NGO) 156, **196** *gloss.*
North Atlantic Treaty Organization (NATO) 21, 22, 88, 90, 93, 143, 145, 151, 152, 156, **195** *gloss.*

Officer Commanding (OC) 33, 88, 179, 190, **196** *gloss.*
Operation Anaconda 27, 29
Operation Apollo 1, 13, **19** *notes*, 21, 39, 41-46, 48-52, 55, **63** *notes*, **64** *notes*, 65, **76** *notes*, 95, 97, 98
Operation Athena 79, 83, 95, 98, 115, 131, 189, 190, 191
Operation Cherokee Sky 34
Operation Echo 39, 41, 43, 44, 46, 48-52
Operation Friction 178
Operation Harmony 179, 180
Operation Horseshoe 146
Operation Kinetic 83, 84
Operation Medusa 169
Operation Palladium 84
Orders group (O-group) 1, 3, 5, 138, **196** *gloss.*
Ottawa **xxiv** *notes*, **xxv** *notes*, **xxvii** *notes*, 4, **19** *notes*, 55, 56, 60, **63** *notes*, **76** *notes*, 179, 189, 190
Otter, William xiv, xv, **xxiv** *notes*, **xxv** *notes*

Pagman 156
Pakistan 30, 155
Peacetime 39, 176
Peers 42, 73, 74, 97, 100, 118
Persian Gulf 13, 55, 65, 83, 104
Personnel Evaluation Reports (PER) 4, **196** *gloss.*
Petawawa 3, 132, 135, 138, 139, 165, 179, 191
Physical Training (PT) 182, **196** *gloss.*
Pilots 39, 43, 44, 48
Poppy Farmers 156
Prelip 87
Prime Minister 24
Princess Patricia's Canadian Light Infantry (PPCLI) 21-23, 26, 121, 123, 129, 165, 191
Pristina Airport 88
Professional Development (PD) iii, v, vi, ix, 129, 149, 150, 162, **196** *gloss.*
Professionalism v, xv, 13, 18, 40, 58, 76, 100
Provincial Reconstruction Team (PRT) 116, 135, 136, 140, 142, 148, 154, 158, 159, 171, **196** *gloss.*

Quality of Life 4, 80, 82, 97, 161
Quick Reaction Force (QRF) 33, 133, 135, 146, 151, 157, **196** *gloss.*

Ramadan 93, 104
Reconnaissance (Recce) 1, 10, 24, 28, 34, 88, 90, 131-136, 138, 139, 141, 142, 144-146, 149, 153, 154, 156-159, 165, 167, 170, 175, **196** *gloss.*
Reconstruction 116, 127, 135, 142, 171, **196** *gloss.*
Regimental Sergeant-Major (RSM) 2, 21, 22, 28, 107, 115, 116, 121-123, 129, 131-133, 135, 138, 148, 190, 191, **197** *gloss.*
Regimental system 176
Repatriate 45, 46, 53
Reporter 29, 36

Reserves **xxvii** *notes*, 159, 173, 175, 180, 181, **186** *notes*, 189
Rest and Relaxation (R&R) 7, 30, 31, 46, 48, 92, 103, 144, 161, **196** *gloss.*
Rocket Attack 80, 82, 109, 135, 146, 147, 153, 156, 157, 171
Rocket Propelled Grenade (RPG) 128, 157, **197** *gloss.*
Royal 22e Regiment (R22eR) 116, 119, 185, **196** *gloss.*
Royal Canadian Regiment (RCR) 115, 139, 168, 190
Rules of Engagement (RoE) 17, 18, 94, 105, 117, **196** *gloss.*
Rumours 59-61, 63, 65
Russian 28, 88, 91, 125, 127, 149
Rwanda xii, 87, 178, 191

Seaman's Club 70, 71
Security iii, v, vi, ix-xi, xix, 17, 21, 24, 26, 27, 32, 39, 41, 44, 50, 53, 66, 79, 80, 82, 86, 91, 96, 101, 111, 115, 117, 126-128, 133, 142, 144, 145, 147, 149, 152, 153, 155, 165, 172, **193** *gloss.*, **195** *gloss.*
Senior leadership 18, 19, 28, 103, 173
Serbia 84, 86-88, 91
Shi A Kot Valley 27
Shura 125
Skopje Airport 87, 89
Somalia x, xii, xvii, **xxv** *notes*, 178
South East Asia xii
South West Asia (SWA) 1, 2, 8, 31, 39-41, 44, 47, 51, 95, 189, 190, **197** *gloss.*
Special Forces (SF) 23, 32, 34, 157
Specialist Engineering Team (SET) 1, 3-7, 10
Strait of Hormuz 55
Strategic Planning Team 158, 159
Strategic Soldier xxi
Suicide Bomber xi, 151-153, 156, 167
Surobi 155

Tactical Airlift Detachment (TAL Det) 39-42, 44, 48, 53, 95, 189, **197** *gloss.*
Taliban 15, 22, 30, 32, 34, 55, 128, 156, 157, 167, 169, 171
Tarnac Farm Facility 31
Task Force (TF) 2, 23, 33, 39, **63** *notes*, 83, 90, 95, 96, 116-118, 131, 133-148, 152-154, 157-162, 165-174, 181, 189, **194** *gloss.*, **195** *gloss.*, **197** *gloss.*
Task Force Aviano 39
Task Force Kabul (TFK) 83, 95, 96, 99-105, 134, 142, **197** *gloss.*
Team 1-3, 6, 7, 10, 13-15, 17, 18, **19** *notes*, **20** *notes*, 22, 25, 29, 30, 32, 34, 35, 41-43, 47-49, 56, 58, 61, 62, 65-67, 69, 70, 73-75, 86, 89, 96, 97, 101, 113, 116, 117, 132, 134, 135, 141, 148-150, 157-159, 171, 183, **187** *notes*
Team building 14, **20** *notes*, 25, 41, 56, 75, 86, 96, 97
Technician xviii, 3, 41, 51, 107, **186** *notes*, 189, 190
Tempo xvi, 46, 82, 162, 173, 184
Terms of Reference 81, **197** *gloss.*
Terrorism 18, **19** *notes*, 21, 23, **63** *notes*, **76** *notes*, 95, 110
Theatre v, 2, 3, 6, 10, 15, 22, 31, 33, 40, 42, 46, 47, 57-60, **63** *notes*, 68-71, 74, **76** *notes*, 81, 83, 84, 95, 101, 108, 110, 111, 115, 117-119, 122, 123, 125, 131, 134-137, 139, 140, 143, 145, 146, 148, 152-154, 158, 160-163, 166, 167, 169, 172, **197** *gloss.*
Theatre Support Element (TSE) 83, 96, 98, 100, 101, 134, 137, 141, 142, 148, 159, 160, 167, **197** *gloss.*

Town Hall meeting 137, 148
Training iii, ix, x, xiv-xviii, xxi, xxii, 1, 13-15, 17, 18, 23, 24, 26, 31, 32, 67-69, 81, 82, 86, 90, 91, 104, 110-113, 117, 118, 122-124, 126-128, 133-140, 146, 147, 150, 155, 161-163, 166, 169-172, 175, 177-182, 184, **186** *notes*, 191, **195** *gloss.*, **196** *gloss.*, **197** *gloss.*
Transfer of Command Authority (TOCA) 136, 158, **197** *gloss.*
Trust xiii, 14, 27, 56, 61, 62, 71, 113, 119, 125, 126, 148, 182-184, **186** *notes*
Tuckman, Bruce 182, **186** *notes*

United Arab Emirates (UAE) 30, 31, 134, **197** *gloss.*
United Nations (UN) xxiii, 87, 142, 154, **197** *gloss.*
Unknown xxii, 14, 32, 33, 70, 123, 124, 126, 132, 144, 145, 155, 156, 176
US xi, xii, 1, 15, 24-27, 30, 111, 134, 156
US Air Force 25
US Forces 134

van Riper, Paul xii, **xxiv** *notes*
Vice-Chief of the Defence Staff 70
von Steuben, Frederick xiii

War on Terror xi, 13, 24, 25, 55
Welfare xii, 19, 31, 40, 45, 62, 63, 71, 75, 92, 93, 103, 104, 122, 180, **187** *notes*
White Paper 178

Yugoslavia x, 178, 179

Zahari-Panjwayi 166